Solid State
Engineering Physics

Solid State
Engineering Physics

Dr. Prabir K Basu

Professor
Department of Electronics
LIMAT, Faridabad

Hrishikesh Dhasmana

Sr. Lecturer
Applied Science and Humanities
LIMAT, Faridabad

Taylor & Francis
Taylor & Francis Group
Boca Raton London New York

CRC is an imprint of the Taylor & Francis Group,
an informa business

Ane Books India

Solid State Engineering Physics

© **Reserved**

First Published in 2009 by

Ane Books India

4821 Parwana Bhawan, 1st Floor
24 Ansari Road, Darya Ganj, New Delhi -110 002, India
Tel: +91 (011) 2327 6843-44, 2324 6385
Fax: +91 (011) 2327 6863
e-mail: kapoor@anebooks.com
Website: www.anebooks.com

For

CRC Press
Taylor & Francis Group
6000 Broken Sound Parkway, NW, Suite 300
Boca Raton, FL 33487 U.S.A.
Tel : 561 998 2541
Fax : 561 997 7249 or 561 998 2559
Web : www.taylorandfrancis.com

For distribution in rest of the world other than the Indian sub-continent

ISBN-10 : 1 43980 647 0
ISBN-13 : 978 1 43980 647 0

British Library Cataloguing in Publication Data
A catalogue record for this book is available from the British Library

Printed at Gopsons Paper Ltd., Noida (U.P.), INDIA

Dedicated to

- ❀ *Our Parents*
- ❀ *Roni, Piu, Saksham*
  and all Our Well-wishers

Preface

We feel great pleasure in bringing out a book on Physics for engineering students. The authors feel the need of writing a book according to the use of engineering students in a simple possible way. As we all know that physics is not only a subject, but it is a realization of phenomena happening around us. So this book is practically based on lecture notes which were applied to teach the students in such a way that it gives them an ample opportunity to understand the nutty gritty of different engineering aspects. Every effort has been made to make all the topics presentable in a lucid manner. The authors have consulted various books, journals, libraries and visited several internet sites to prepare the book. The whole syllabus dealt with several branches of physics applicable to engineering students. The philosophy of presentation and material content in the book are based on concept based approach towards the subject. This gives knowledge of current technology which will be indispensable in 21st century. For example, by carrying the knowledge of this syllabus they can easily get the understanding of nanotechnology, emerging as a latest multidisciplinary subject. We have tried to put hard efforts in placing contents in this book such that students feel comfortable with the syllabus in an interesting way.

We sincerely express our deep sense of gratitude to Prof. G.V.K. Sinha, Chairman, LIMAT, Dr. P. Gadde, LIMAT and Prof. (Dr.) H.B. Singh, HOD, Applied Sciences and Humanities for their inspirational influence during our last years of teaching at LIMAT, Faridabad. This only pushed us to write the book. We are grateful to our colleagues Prof. (Dr.) K.L. Moza, Venkatesh K, Dr. Ratna, Sandeep, Dharmender, Roop Singh, Rahul and Mr. N. Udayakumar, Chairman, Udhaya Energy Photovoltaics, Combatore and other friends for their encouragement during preparation of manuscript. We are thankful to the students for giving correct feedback about our approach in writing the book. We both are also very much thankful to all our family members for their heartiest cooperation during the preparation of the book. We are also grateful to the publishers for timely publication of the book.

It will be kind enough for readers for any suggestions which can easily be accepted for further improvement in next edition.

Dr P.K. Basu
Hrishikesh Dhasmana

Contents

CHAPTER 5   PHOTOCONDUCTIVITY AND PHOTOVOLTAICS

CHAPTER 6   MAGNETIC MATERIALS

1

CRYSTAL STRUCTURE

Human beings are using the materials from past in one or other way to meet their daily needs. Materials have always been an integral part of progress of human civilization. The civilizations have been designated by the level of their material development. According to the need of the society, man discovered techniques for producing materials that had properties to those of the materials occurring naturally. The development of many advanced technologies has made life easier. For example, in automobile and aircraft materials used were mostly metallic. As metal has high density so there was limitation on achieving high speed. Due to the advancement in technologies metals are being replaced by composite materials. Now-a-days vehicles are light weight and can achieve very high speed. To apply the materials in any field we should aware of the internal structure of the materials. This is very important part of the observation of any material in different applications so that the criticalities in achieving various applications of the material can defined clearly. Structure of the materials can be studied at various levels of observations depending on the magnification and resolution of physical aids used. For example, macrostructure of material is examined with naked eye or under a low magnification optical instrument. Same way microstructure of materials is examined under higher magnification. So under higher magnification crystals may be studied. Crystal structure deals with the repetitive atomic arrangement. Based on crystal structure, solids are divided into crystalline and non-crystalline solid. The name crystal comes from the Greek word 'krystallos' which means clear ice. It was first applied to describe beautiful transparent quartz 'stones' found in the Swiss Alps. This chapter starts with crystallography. Crystallography is the study of internal structure of crystals, their properties, and external or internal symmetries of crystals. Various techniques applied in industry for determining crystal structure have been described. It further proceeds with fundamental concepts behind solid formation and the various bonds responsible for formation of solid. As we are dealing with crystal structure it is considered a perfect arrangement of atoms but in normal practice it is not so. Therefore point defects are being taken in both ways, qualitative and quantitative. Finally we have put some light about the elementary particles of material.

1. CRYSTAL STRUCTURE

The intermolecular forces bind a large number of atoms into a form, which has a volume and a definite shape. Such aggregates of atoms, which preserve their volumes and shapes, unless subjected to large external force are called **Solids**.

The solids are generally divided into two categories.

(A) Amorphous : The amorphous solids are those which lack the regular arrangement of atoms or molecules in long range. Examples are pitch, plastics, etc.

(B) Crystalline : The crystalline solids are those which contain the regular and repeated patterns of atoms and molecules, and hence a long range order of atoms and molecules. The examples are shown below :

Diamond crystals
used for abrasives

Quartz crystals

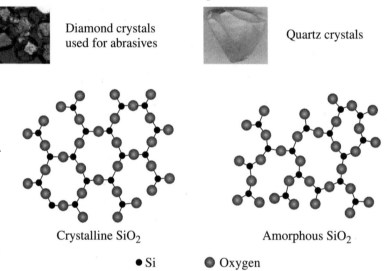

Crystalline SiO$_2$ Amorphous SiO$_2$

● Si   ◉ Oxygen

Fig. 1   Some examples of crystalline and amorphous materials.

Space Lattice

The crystal structure of a solid can be described with the aid of a simple geometrical concept namely a space lattice. The space lattice may be defined as an array of points in space such that the environment about each point is same.

The Basis and Crystal Structure

A crystal lattice is a space lattice in which the lattice sites are occupied by atoms or clusters of atoms. Each lattice point is associated with the same unit of group of atoms, called the basis (or pattern). Thus basis is an aggregate of atoms occupying space point's position. When the basis is repeated with correct periodicity in all directions it gives the actual Crystal Structure. The crystal structure is real, whereas, the lattice is imaginary.

Hence,   space lattice + basis = crystal structure

Lattice does not have any physical significance. It is just a convenient way of representation of position of atoms in a crystal by points. In two dimensions it is called a plane lattice, while in three dimensions it is known as space lattice.

(i) Plane lattice : It may be defined as an infinite array of points in two dimensions in such a way that each point has identical environment. There are generally 5 lattices in two dimensions as given below: 1. Oblique, 2. Square, 3. Hexagonal, 4. Rectangular, and 5. Centered rectangular.

(ii) Space lattice : In 1848 A. Bravais, the French crystallographer proved that there are only 14 space lattices in total, which are required to describe all possible arrangements of points in space. It is subjected to the condition that each lattice point has exactly identical environment. The 14 space lattices are called as Bravais lattice and are shown in figure (2).

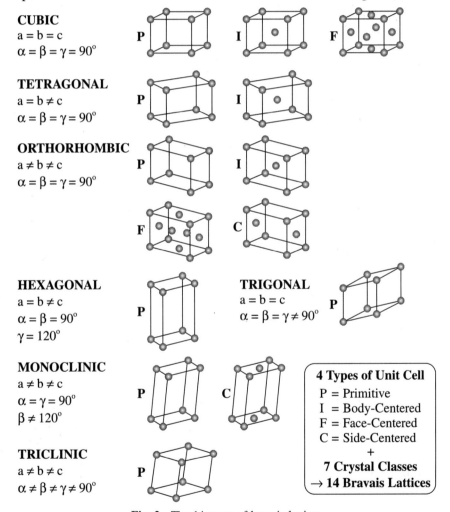

CUBIC
a = b = c
$\alpha = \beta = \gamma = 90°$

TETRAGONAL
a = b ≠ c
$\alpha = \beta = \gamma = 90°$

ORTHORHOMBIC
a ≠ b ≠ c
$\alpha = \beta = \gamma = 90°$

HEXAGONAL
a = b ≠ c
$\alpha = \beta = 90°$
$\gamma = 120°$

TRIGONAL
a = b = c
$\alpha = \beta = \gamma \neq 90°$

MONOCLINIC
a ≠ b ≠ c
$\alpha = \gamma = 90°$
$\beta \neq 120°$

TRICLINIC
a ≠ b ≠ c
$\alpha \neq \beta \neq \gamma \neq 90°$

4 Types of Unit Cell
P = Primitive
I = Body-Centered
F = Face-Centered
C = Side-Centered
+
7 Crystal Classes
→ **14 Bravais Lattices**

Fig. 2 The 14 types of bravais lattices.

The examples of seven crystal systems and parameters of the corresponding unit cell are given below :

1. **Cubic :** Au, Cu, NaCl,

2. **Tetragonal :** SiO_2, TiO_2,

3. **Orthorhombic :** KNO_3, $MgSO_4$,

4. **Trigonal (Rhombohedron) :** As, Sb, Bi,

5. **Hexagonal :** SiO_2, Zn, Cd, Mg,

6. **Monoclinic :** $CaSO_4$, $2H_2O$, $FeSO_4$, Na_2SO_4,

7. **Triclinic :** $K_2Cr_2O_7$

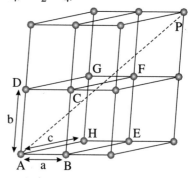

Unit Cell and Translational Vectors : In the three dimensional figure 3 below we have shown a part of the crystal. It has a three dimensional unit cell *ABCDEFGH* as a cuboid. Its sides are $AB = a$, $AD = b$ & $AH = c$, then, by rotating this cuboid by any integral multiple of vectors, a, b and c the whole crystal lattice is obtained. This way the fundamental unit ABCDEFGH is called a unit cell. The dotted line *AP* represents the translational vector.

Fig. 3   A part of a three dimensional crystal with its atomic arrangement.

- A unit cell is the smallest geometrical figure, the repetition of which gives the actual crystal structure. It is the fundamental elementary pattern of minimum. no. of atoms, molecules, which represent fully all the characteristics of the crystal.

- The periodically repeating arrangement of points in a space lattice can be described by the operation of parallel displacement, called a 'Translation'.

The position of any lattice point *D* with respect to another lattice point *A* in the crystal is subjected to the three different translational $\vec{a}$, $\vec{b}$, & $\vec{c}$ along *x, y,* & *z* directions respectively and is given by

$$\vec{AP} = \vec{l} = p\,\vec{a} + q\,\vec{b} + r\,\vec{c};$$ where, *p, q* & *r* are integers.

The vectors $\vec{a}$, $\vec{b}$ & $\vec{c}$ **are translational vectors, also known as basis vectors or primitive vectors.**

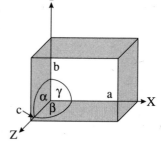

Fig. 4   A three dimensional unit cell.

A three dimensional unit cell is shown in figure 4 aside. The unit cell is a parallelepiped formed by translational $\vec{a}$, $\vec{b}$ & $\vec{c}$ vectors as concurrent edges.

Miller Indices

The Crystal lattice may be regarded as made up of an aggregate of a set of parallel equidistant planes passing through lattice points. These planes are known as **lattice planes**. For a given lattice, planes can be chosen in a different number of ways (as shown in figure 5 below) to evolve a difficulty to designate them. In the figure 5, there are many planes passing through

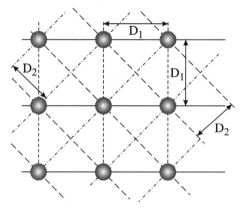

Fig. 5 Two dimensional array of atoms in a crystal with different possible lattice planes.

same nine atoms with different plane lattice with different interplaner spacing D1 and D2. One scientist, Miller, developed a procedure by using three numbers, (*hkl*), known as **Miller Indices**, to designate the planes. He introduced the way to represent the position of all types of planes in crystal structure within the unit cell. Miller planes are represented by (*hkl*) and their family is by {*hkl*}. Direction of Miller planes are represented by <*hkl*> and their family by [*hkl*]. This means that (110) & ($\overline{1}10$) planes correspond to same family of planes {110}. Similarly <110> & <$\overline{1}\overline{1}$0> directions correspond to same family of directions [110].

- The Miller Indices are the set of three smallest possible integers which have the same ratios as the reciprocals of the intercepts of the plane concerned on the three crystallographic axes.

A. Rules to Find Miller Indices :

Step-1 : Determine the intercepts of plane on three coordinate axes.

Step-2 : Take reciprocals of these.

Step-3 : Reduce the reciprocals into whole number by LCM.

Example : From the figure 6 below we have

Step-1 : Intercepts are 3, 2, 1

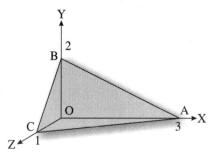

Fig. 6 A crystal plane *ABC* intercepting the axes.

Step-2 : Their reciprocals are 1/3, 1/2, 1

Step-3 : LCM of denominator of the reciprocals is 6.

Hence multiplying all the three reciprocals by 6, we get, 2, 3, 6. Hence the **Miller Indices** here are **(236)**.

B. Drawing of Lattice plane using Miller Indices : Let the Miller Indices of the plane is given as (123).

Step-1 : Its reciprocals are 1, 1/2, 1/3.

Step-2 : In coordinate axes as shown in figure 7 aside we mark a unit of magnitude 1 (as maximum possible in the drawing) in all X, Y and Z-axis.

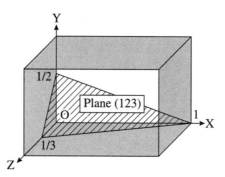

Fig. 7 (123) lattice plane in a unit cell.

Step-3 : Now, as for our plane we mark the reciprocals as intercepts on the axes as in figure 7 as points A, B and C respectively. Now we join ABC. This is our required plane.

Some more planes are shown in the following figures (8).

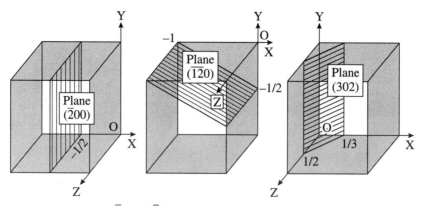

Fig. 8 ($\bar{2}$00), (1$\bar{2}$0) and (302) lattice planes in a unit cell.

I. Some concepts of Crystal Structure

(A) Separation between Lattice Planes : We consider a plane ABC of a cubic crystal as shown in figure 9(a) given below :

Its Miller Indices are (*hkl*). *ON* is the perpendicular drawn from the origin to this plane. This distance *ON* represents the **interplaner distance** **'d'** of the family of planes with one plane at the origin. Let α, β and γ are the angles between coordinate axes *X, Y, Z* and *ON* respectively as shown in Figure 9(b).

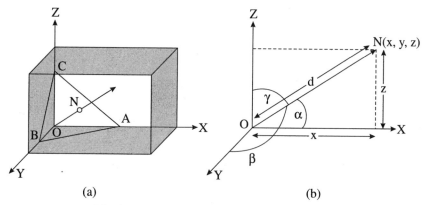

(a)   (b)

Fig. 9 (a) Lattice place *ABC* inside the unit cell,
(b) It shows the angles made by interplaner distance *ON* with the three axes.

So,
$$OA = \frac{a}{h}, \; OB = \frac{a}{k}, \; OC = \frac{a}{l} \qquad \qquad ...(1)$$

Because, if *OA*, *OB* and *OC* are intercepts, then $h = \left(\frac{1}{OA}\right)a$, $k = \left(\frac{1}{OB}\right)a$; $l = \left(\frac{1}{OC}\right)a$; where, *a* is the LCM of *OA*, *OB* and *OC* and also the length of the cube edge as shown in Figure 9(a).

Also, from Figure 9(b) before,

$$\cos \alpha = \frac{ON}{OA} = \frac{d}{OA},$$

$$\cos \beta = \frac{d}{OB},$$

and
$$\cos \gamma = \frac{d}{OC} \qquad \qquad ...(2)$$

From Figure 9(b),

$$ON = \sqrt{x^2 + y^2 + z^2} \qquad \qquad ...(3)$$

and
$$\frac{x}{d} = \cos \alpha$$

$$\Rightarrow \qquad \qquad x = d \cos \alpha$$

Similarly, $\qquad y = d \cos \beta \; ; \; z = d \cos \gamma$ ...(4)

Using eqns. (3) and (4)

$$d = \sqrt{[(d \cos \alpha)^2 + (d \cos \beta)^2 + (d \cos \gamma)^2]}$$

$\Rightarrow \qquad \cos^2 \alpha + \cos^2 \beta + \cos^2 \gamma = 1$

$\Rightarrow \qquad \left(\dfrac{d}{OA}\right)^2 + \left(\dfrac{d}{OB}\right)^2 + \left(\dfrac{d}{OC}\right)^2 = 1$

Using eqn. (2) we have

$$\left[\frac{d^2 \, (h^2 + k^2 + l^2)}{a^2}\right] = 1$$

$\Rightarrow \qquad d = \dfrac{a}{\sqrt{h^2 + k^2 + l^2}}$ ...(5)

This is the relation between d & a by using (hkl).

II. Values of d for Different Planes with Simple Cubic (SCC) and Face Centered (FCC) Structures

(a) **SCC :** Here, the (100) planes cut X-axis and are parallel to Y and Z-axis.

The (110) planes cut obliquely across X & Y-axis and are parallel to Z-axis.

The (111) planes cut obliquely across X, Y & Z-axis.

Hence, $\qquad d_{100} = \dfrac{a}{\sqrt{h^2 + k^2 + l^2}}$

$\qquad\qquad = \dfrac{a}{\sqrt{1^2 + 0^2 + 0^2}} = a \; ;$

$\qquad d_{110} = \dfrac{a}{\sqrt{1^2 + 1^2 + 0^2}} = a/\sqrt{2} \; ;$

$\qquad d_{111} = \dfrac{a}{\sqrt{1^2 + 1^2 + 1^2}} = a/\sqrt{3}$

Hence, $\qquad d_{100} : d_{110} : d_{111} = 1 : 1/\sqrt{2} : 1/\sqrt{3}$

(b) **FCC :** It is clear from Figure 10 next that the additional (100) planes arise halfway between original (100) planes, which are due to atoms placed on faces of each unit cell. Hence, the distance between d_{100} planes is a/2.

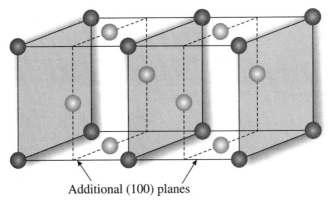

Additional (100) planes

Fig. 10 Additional (100) planes in FCC crystal.

For (110) planes as in Figure 11 next, we find that an additional plane of (110) comes midway between the planes of original (110) planes. So the distance between (110) planes are half of that in simple lattice, *i.e.*,

$$d_{110} = \frac{1}{2} d_{110} \bigg|_{FCC} = \frac{1}{2} \frac{a}{\sqrt{2}} = \frac{a}{2\sqrt{2}}$$

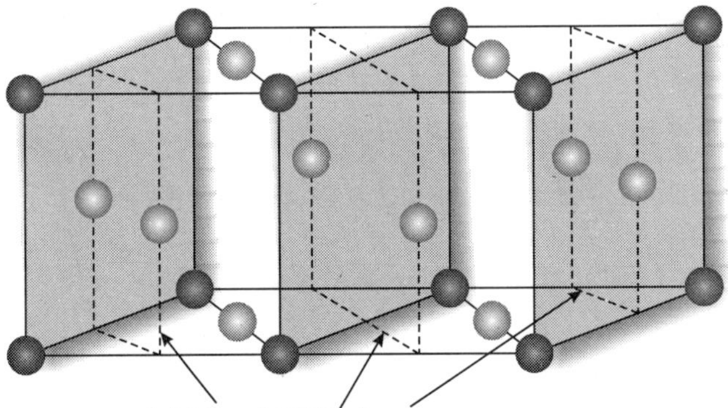

Additional (110) planes

Fig. 11 Additional (110) planes in FCC crystal.

For (111) planes, as in Figure 12 below, we find that no additional plane arise.

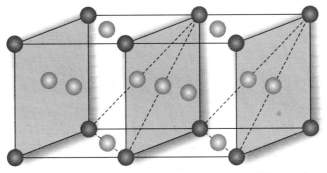

Fig. 12 No additional (111) plane possible in FCC crystal.

Hence, $d_{111} = a/\sqrt{3}$

Finally, for FCC crystals,

$$d_{100} : d_{110} : d_{111} = 1 : 1/\sqrt{2} : 2/3\sqrt{3}$$

(c) BCC : It is clear from Figure 13 below that due to the presence of an additional point at the body centre, the interplaner spacing for three low

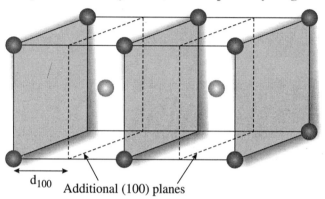

Fig. 13   Additional (100) planes in BCC crystal.

index planes (100), (110) and (111) give slightly different results. There appears an additional plane halfway between (100) and (111) planes.

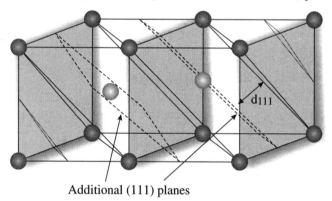

Fig. 14   Additional (111) planes in BCC crystal.

Due to the presence of an additional plane at the body centre, the interplaner spacing for three low index planes (100), (110) and (111) give slightly different result. There appears an additional plane half way between (100) and (111) planes as shown in Figure 14 and Figure 15.

Here, distance between (100) planes is half of that in simple lattice, *i.e.*,

$$d_{100} = \frac{1}{2} d_{100}\Big|_{SCC} = \frac{1}{2} a.$$

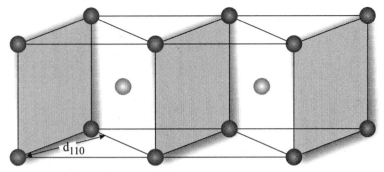

No additional (110) planes

Fig. 15 No additional (110) plane possible for BCC crystal.

However, distance between (110) planes is same as in simple lattice,

i.e.,
$$d_{110} = d_{110}\big|_{SCC} = \frac{a}{\sqrt{2}}$$

as there is no additional plane as shown in Figure 15.

The distance between (111) planes is half of that in simple lattice,

i.e.,
$$d_{111} = \frac{1}{2} d_{111}\bigg|_{SCC} = \frac{a}{2\sqrt{3}}.$$

Finally, for BCC crystals,

$$d_{100} : d_{110} : d_{111} = 1 : \sqrt{2} : 1/\sqrt{3}$$

III. Some Relevant Physical Parameters

1. Co-ordination Number (CN) : Co-ordination number is the number of nearest neighbours at equal distance to an atom in a given structure. These are shown in Figure (16).

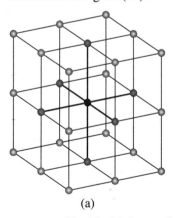

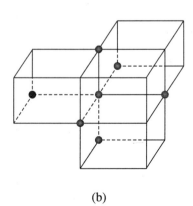

| (a) | (b) |

Fig. 16 (a) A part of SCC crystal showing nearest neighbours
(b) Same drawing only showing four adjacent unit cells.

For SCC : CN = 6 as each atom is shared by 6 atoms with a distance between neighbouring atoms is 'a'. This is shown in Figure 16.

For BCC : CN = 8 as each corner atom of unit cell is at a distance, $\frac{\sqrt{3}}{2} a$ to body centered atom. Body centered atom is represented by "**O**" in the Figure 17.

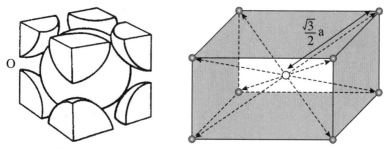

Fig. 17   A unit cell of BCC crystal showing its nearest neighbours.

For FCC : CN = 12 as four corner atoms are at a distance, $a/\sqrt{2}$ to face centered atoms and eight face centered atoms are also at a distance $a/\sqrt{2}$. Face centered atoms are represented by "**O**" in the Figure 18.

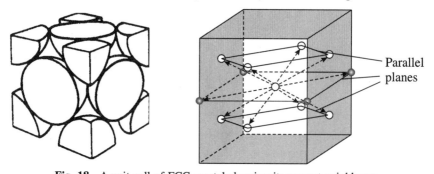

Fig. 18   A unit cell of FCC crystal showing its nearest neighbours.

2. Number of atoms in a Unit Cell (N) : This is given by

$$N = N_i + \frac{N_f}{2} + \frac{N_C}{8}.$$

Here, N_i is no. of interior or body centered atoms, N_f is no. of face centered atoms and N_C is no. of corner atoms.

For SCC : $N_i = 0, N_f = 0, N_C = 8$;   Therefore, $N = 1$.
For FCC : $N_i = 0, N_f = 6, N_C = 8$;   Therefore, $N = 4$.
For BCC : $N_i = 1, N_f = 0, N_C = 8$;   Therefore, $N = 2$.

3. Packing Fraction (APF) : The fraction of space occupied by atoms in a unit cell is known as the atomic packing fraction. It is defined as the fraction of the volume occupied by the atoms in the unit cell to the total volume of the unit cell. In figures below R is the radius of the atom.

(a) For SCC :

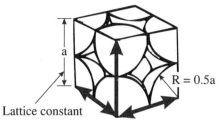

Fig. 19   An SCC unit cell with maximum possible packing.

APF for SCC $= \dfrac{\pi}{6} = 0.52$

(b) For BCC : Here $N = 2$ and one atom is at the centre (as shown in Figure 20 below) :

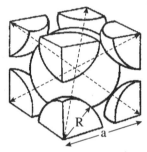

Fig. 20   A BCC unit cell with maximum possible packing.

APF for BCC $= \dfrac{\pi\sqrt{2}}{6} = 0.68$

Close-packed directions :

$$\text{length} = \text{body diagonal} = 4R = \sqrt{3}\,a$$

(c) For FCC : Here, $N = 4$.

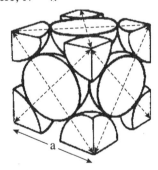

Fig. 21   An FCC unit cell with maximum possible packing.

APF for FCC $= \dfrac{\pi}{3\sqrt{2}} = 0.74$

Close-packed directions :

$$\text{length} = \text{face diagonal} = 4R = \sqrt{2}\,a$$

4. Nearest Neighbour Distance : This is the distance between the centers of two neighbouring atoms in contact.

5. Atomic Radius : This is the half of the nearest neighbour distance.

6. Density : The density of any material is the mass per unit volume. For unit cell also this definition works and the density of the unit cell is same as the density of the crystal itself. Hence, density, ρ, is given by

$$\rho = \frac{\text{Mass}}{\text{Volume}} = \frac{N \times M}{V N_A}$$

Here, N, M, V, N_A are number of atom per unit cell, atomic weight volume of atom and Avagadro's number respectively.

IV. Sodium Chloride Structure

Ionic solids are made up of cation and anions. The cations (+ve) are smaller in size to the anions as cations have given up their valence electrons to anions. The anions form the closed pack structure into whose voids cations get accommodated as shown in Figure 22. Each cation prefers to have as

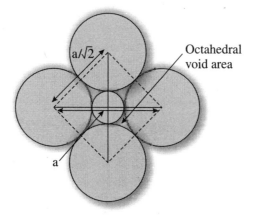

Fig. 22   NaCl packing in one face of the crystal.

many nearest neighbours as possible. So does the anion. Therefore packing arrangements are possible due to different sizes of the ions. In NaCl structure shown in Figure 23 below, it is like the interpenetration of two FCC lattices. One is sodium ion lattice & the other chlorine ion lattice. It is having a co-ordination no. 6, so 6 anions surround each cation and 6 cations surround

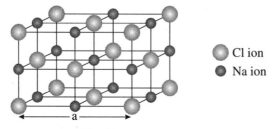

○ Cl ion
● Na ion

Fig. 23   NaCl structure.

each anion. The cations occupy the void formed by the ions. As there is one octahedral void, so 6 anions surround each cation and 6 cations surround each anion. The cations occupy the void per anion (Na : Cl = 1 : 1), so all octahedral voids in this structure are occupied. In the unit cell, the 8 chlorine ions are in the corners shared by 8 unit cells around and 6 ions at the faces each of which is shared by 2 conjugated cells.

Hence, total no. of chlorine ions per unit cell (*i.e.*, for FCC) $= 0 + 8 \times \dfrac{1}{8} + 6 \times \dfrac{1}{2} = 4$. Similarly, 12 sodium ions located at the edges of the cube & each sodium ion is shared by 4 conjugated unit cells. One sodium ion is located at the centre of the unit cell. Hence, total no. of sodium ions

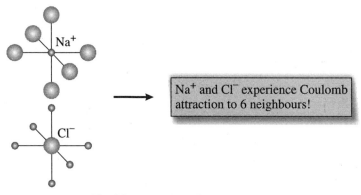

Fig. 24   Formation of NaCl structure.

per unit cell (*i.e.*, for FCC) $= 0 + 12 \times \dfrac{1}{4} + 8 \times \dfrac{1}{8} = 4$. Thus the total nos. of sodium and chlorine ions per unit cell is 8 and hence there are 4 sodium chloride molecules per unit cell. The formation of NaCl molecule is shown in Fig. 24.

2. | BONDING IN SOLIDS

The existence of matter to be in solid state lies in the interatomic forces. This force comes into play when the atoms are brought in close to each other. For a solid we have a solid structure.

I. Different Forces in a Crystals

There exist two types of forces in the crystal.

(a) Attractive force : This is between constituent particles (atoms or molecules) which keep them together and prevent the atoms moving away from each other. The Potential Energy (PE) due to this force is negative because the atoms do work of attraction and is given by

$$U_1 = -\frac{A}{r^m}$$

(b) Repulsive force : This is between constituent particles (atoms or molecules) since huge force is required to compress a solid to a appreciable extent. The Potential Energy due to this force is positive because external work must be done to bring two such atoms close together so that they repel each other and is given by

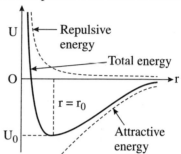

$$U_2 = \frac{B}{r^n}.$$

Here, A, B are constants; m, n are integers; r is interatomic or intermolecular distance and is short range in nature which prevents the crystal from collapsing indefinitely and is also the characteristics the molecule or compound.

Fig. 25 The variation of total potential energy with interatomic distance.

Hence, the net (Total) PE is given by

$$U = U_1 + U_2 = -\frac{A}{r^m} + \frac{B}{r^n}$$

- At large r, $U_1 \to 0$, but decreases gradually at small r.
- With a decrease in r, U_2 increases rapidly, since positive charges on nuclei repel each other very strongly as r decreases. The force between atoms can be expressed as :

$$F = -\frac{dU}{dr} = -\frac{mA}{r^{m+1}} + \frac{nB}{r^{n+1}}.$$

From the above Figure 25 we see that at large distances the atoms do not interact each other, so $U = 0$. With a decrease in interatomic distances force is attractive due to positive and negative charges of atoms and becomes zero when this separation becomes comparable to the atomic diameter. When the distance is further decreased the force becomes repulsive so that the reduction in interatomic distance can not take place. So there is a compromise between the above two opposite forces at $r = r_0$. This corresponds to stable equilibrium configuration. At this separation U is the minimum and force is zero.

The corresponding $U(r_0)$ is negative and is known as **cohesive energy** or **dissociation energy** of the molecule

$\Rightarrow$ Energy required to separate the two atoms of the solid.

At $r = r_0$, net force is zero, *i.e.*,

$$F = \left(\frac{dU}{dr}\right)_{r = r_0} = 0$$

$\Rightarrow$

$$\frac{mA}{r_0^{m+1}} = \frac{nB}{r_0^{n+1}}$$

$\Rightarrow$

$$r_0 = \left(\frac{nB}{mA}\right)^{1/n-m}$$

The energy corresponding to the equilibrium state can be obtained as

$$U_0 = -\frac{A}{r_0^m}\left(1 - \frac{m}{n}\right).$$

From above equation it is clear that minimum in the energy curve is possible only if $n > m$. This may be shown by employing equilibrium condition for U as

$$\left(\frac{d^2U}{dr^2}\right)_{r=r_0} > 0$$

$$\Rightarrow \qquad -\frac{m(m+1)}{r_0^{m+2}} + \frac{n(n+1)}{r_0^{n+2}} > 0$$

Substituting the value of r_0 we can obtain the condition of $n > m$ at equilibrium position.

II. Bonding

Based on the bond strength, atomic bonding can be grouped into primary and secondary bonding. Primary bondings have bond energies in the range of $0.1 - 10$ eV/bond and act between atom to atom. Ionic, covalent and metallic bondings are examples of primary bonding. However, secondary bonding have energies in the range of $0.01 - 0.5$ eV/bond. This bond acts between molecule to molecules and comparatively two orders of magnitude weaker than primary bond. Hydrogen bond and Van der Waals bonding are examples of secondary bonding. The crystals may be classified by a variety of ways, *e.g.*, mechanically, electrically and chemically. However, the most convenient way is on the basis of Interatomic binding forces.

According to this scheme the crystals may be categorized into 6 major categories :

(1) Ionic,   (2) Covalent,   (3) Metallic,

(4) Molecular   (5) Hydrogen bonded,

(6) Mixed or Multiple bonded Crystal.

1. Ionic Crystals (NaCl, CsCl, etc) : Here electrons are transferred from one type of atoms (who loses electrons readily) to other atoms (which have high affinity to electrons). So the crystal consists of positively and negatively charged ions.

Example : NaCl : Na is electropositive and Cl is electronegative. The attractive force is the Coulomb attraction between ions and repulsive force is due to overlapping of electronic shells of neighbouring atoms.

Properties :

- Strong bonding, *i.e.*, high melting point (NaCl ≈ 800°C),
- Close packed structure,
- Good conductors as ions, but bad conductors of heat & electricity,
- Transparent over wide range of electromagnetic spectrum.

2. Covalent Crystal (C, Si, Ge) : Here, two similar or identical atoms are brought together to such an interatomic distance that the orbital of unpaired electron of one atom begins to overlap with that of one in other atom. The electrons lying between the two atoms belong equally to both and serve to complete the outer shell of both atoms. Since these sharing electrons have anti-parallel spin, the atoms attract each other and bonding result. This bonding is known as covalent bonding. Here, the valence electrons are shared equally between neighbouring atoms.

Properties :
- Very strongly bound and bonds are directional,
- Very high melting point (*e.g.*, Si ≈ 1412°C),
- Semiconductors,
- Transparent in IR region, *i.e.*, opaque at high frequencies.

3. Metallic Crystal (Na, Al, etc.) : The metallic crystal can be considered as a framework of positive ions surrounded by electron clouds. The crystal is held together by the electrostatic attraction between negative electron gas and positive metal ions. Here, the valence electrons are weakly coupled to nucleus. The cohesion results from a combination of
- the attraction between positive ion cores and electron cloud,
- the mutual repulsion of electrons,
- the mutual repulsion of positive ion cores.

Properties :
- Moderate to Strong bonding,
- Close packed structure,
- Low melting point (Na ≈ 100°C),
- Very good conductors of heat and electricity,
- Ductile,
- Good reflectors,
- Opaque to electromagnetic radiation.

4. Molecular Crystal : It consists of neutral atoms or molecules bound together in the solid phase by weak short range attractive force (Van der Waals force). It depends on the mutual deformation of the atoms which is due to permanent dipole of the molecule on the dipole induced in neighbouring molecules. This leads on the average to an attraction. Van der Waals force is responsible for the condensation of gases into liquids and freezing of liquids into solids in the absence of ionic, covalent or metallic bonding mechanisms.

Properties :
- short range, weak
- responsible for condensation (gas-liquid)
- responsible for freezing (liquid-solid) if no stronger forces present

Reason : temporary fluctuating dipole fields!

Properties :
- Weak binding,
- Transparent,
- Soft, lowest melting,
- Low electrical and thermal conductivity.

5. Hydrogen Bonded Crystals (Ice (H_2O)) : Here the atom of hydrogen is attracted by an extremely electro-negative atom such as O, F, N, Cl to form H-bond. The electronegative ion attracts the bonding electron and becomes negatively charged, the H-atom then assumes a positive charge. The H-bond is a result of electrostatic attraction of these charges.

- Due to Van der Waals force between H containing molecules
- Due to permanent dipole moments
- Occurs in molecules where H is attached to electronegative atoms (*e.g.*, H_2O, NH_3, HCl)
- The electronegative atoms need to have "lone pairs"

This is shown in figure 26.

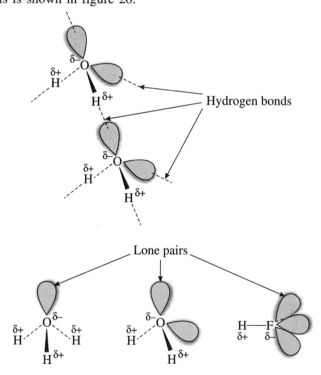

Fig. 26 Formation of hydrogen bonded crystals.

Properties :
- Moderately weak bonding,
- Loose structures,
- Low electrical and thermal conductivities,
- Transparent.

6. Mixed or Multiple bond Crystal : This consists of a mixture of Van der Waals and Ionic Bonding. Examples are salts based on : Tetra-Cyano-Quino-Dimethane TCNQ, N-methyl-Phenazinium NMP, etc.

Properties :

- Range from metallic to insulating.

III. Lattice Energy (Binding Energy) of Ionic Crystal

Lattice energy of an ionic crystal is the energy that will be released by the formation of the crystal from individual neutral atoms. The bond energy is however different than the lattice energy. The stability of an ionic crystal depends on the balancing of at least **three forces :**

- (i) **the electrostatic forces (Coulomb forces)** between ions which are of a attraction falling off with square of the distance,
- (ii) **the Van der Waals force of attraction** falls as seventh power of distance, and
- (iii) **interatomic repulsive forces** fall much more rapidly with distance.

The resultant of the attractive and repulsive forces leads to an equilibrium position of minimum potential energy as in Figure 27 below. For

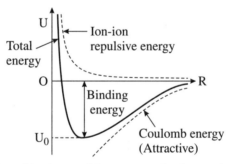

Fig. 27   Variation of ion-ion repulsive energy, columb attractive energy and net energy with interatomic distance for an ionic crystal.

two ions of charges Z_1e & Z_2e separated by a distance R the attractive energy is $= -\dfrac{Z_1Z_2e^2}{4\pi R\varepsilon_0}$, here, Z_1 & Z_2 are valencies of ions. So, for the whole crystal, the Coulomb potential energy (attractive) is

$$U_{\text{attr}} = -\frac{AZ_1Z_2e^2}{4\pi R\varepsilon_0}.$$

This is the net Coulomb potential energy of any one ion due to the presence of all the other ions in the crystal. Here, A is known as **Madelung Constant**.

To stop lattice collapse there are repulsive forces between the ions and becomes appreciable when the electronic shells of neighbouring ions begin to overlap. It was assumed that the repulsive energy due to overlap of the outer electron shells between two ions is inversely proportional to some power of the distance R, i.e., $\dfrac{B}{R^n}$.

Hence, the repulsive energy of a particular ion due to the presence of all other ions in the crystal is $\dfrac{B}{R^n}$, where, n is called repulsive exponent and B is a constant.

Hence, the total energy of one ion due to the presence of all the other ions is

$$U(R) = U_{\text{attr}} + U_{\text{rep}}$$

$$\Rightarrow \qquad U(R) = -\frac{AZ_1Z_2e^2}{4\pi R\varepsilon_0} + \frac{B}{R^n} .$$

For monovalent alkali halides, $Z_1 = Z_2 = 1$.

So, $\qquad U(R) = U_{\text{attr}} + U_{\text{rep}}$

$$\Rightarrow \qquad U(R) = -\frac{Ae^2}{4\pi R\varepsilon_0} + \frac{B}{R^n} .$$

If N_A is Avogadro's number, the total energy of the crystal is

$$U_{\text{Total}} = N_A\, U(R)$$

$$\Rightarrow \qquad U_{\text{Total}} = N_A\left[-\frac{Ae^2}{4\pi\varepsilon_0 R} + \frac{B}{R^n} \right] \text{per Kmol} \qquad \qquad \ldots(1)$$

The energy will be minimum at equilibrium, i.e., at $R = R_0$, where, R_0 is the equilibrium distance,

hence, $\qquad \qquad \left. \dfrac{dU}{dR} \right|_{R=R_0} = 0$

$$\Rightarrow \qquad \left[\frac{Ae^2}{4\pi\varepsilon_0 R_0^2} - \frac{nB}{R_0^{n+1}} \right] = 0$$

$$\Rightarrow \qquad \qquad B = \frac{Ae^2 R_0^{n-1}}{4\pi\varepsilon_0 n} \qquad \qquad \ldots(2)$$

Hence, from eqns. (1) and (2), we have,

$$U_{\text{Total}}\Big|_{R=R_0} = N_A\left[-\frac{Ae^2}{4\pi\varepsilon_0 R_0} + \frac{Ae^2 R_0^{n-1}}{4\pi\varepsilon_0 n R_0^n} \right]$$

$$\Rightarrow \qquad U_{\text{Total}}\Big|_{R=R_0} = N_A \frac{Ae^2}{4\pi\varepsilon_0}\left[-\frac{1}{R_0} + \frac{1}{nR_0} \right]$$

$$\Rightarrow \qquad U_{\text{Total}}\Big|_{R=R_0} = -\frac{N_A\, Ae^2}{4\pi\varepsilon_0 R_0}\left[\frac{n-1}{n} \right] .$$

This is the **Lattice Energy or Binding energy**. For NaCl, $U_{\text{Total}} = 7.95$ eV.

IV. Bond Energy

At equilibrium position, *i.e.*, at $R = R_0$, the energy released in the formation of ionic crystal molecule is called the **bond energy** of the molecule.

So, bond energy,
$$V = -\frac{Z_1 Z_2 \, e^2}{4\pi\varepsilon_0 R_0} \, ;$$

For NaCl, $V = -6$ eV with $R_0 = 0.24$ nm.

Example : For NaCl, energy required to remove one electron from Na-atom outer shell (*i.e.*, the Ionization energy) is 5.1 eV, *i.e.*,

$$\textbf{Na} + \textbf{5.1 eV} \rightarrow \textbf{Na}^+ + e^-$$

The electron affinity of Cl is 3.6 eV. So,

$$\textbf{Cl} + e^- \rightarrow \textbf{Cl}^- + \textbf{3.6 eV}$$

So the net energy spent $= 5.1 - 3.6$ eV $= 1.5$ eV, thus,

$$\textbf{Na} + \textbf{Cl} + \textbf{1.4 eV} \rightarrow \textbf{Na}^+ + \textbf{Cl}^-$$

Now, the bond energy of NaCl $= -6$ eV.

Hence, the energy released in the formation of NaCl molecule starting from Na and Cl atoms having zero point energy is $6 - 1.5$ eV $= 4.5$ eV.

So, conversely, 4.5 eV of energy is required to dissociate NaCl molecule into Na and Cl ions.

3. | CRYSTAL STRUCTURE ANALYSIS

Much of our knowledge of the internal structure of crystals has been obtained from *X*-rays. Diffraction of waves occurs when the waves are scattered by a periodic arrangements of scattering objects separated by distances of the order of wavelengths. A plane diffraction grating is used commonly to study the diffraction of light.

I. Bragg's Law

The dimension of atoms and the interatomic spacing in a crystal are of the order of the wavelength of *X*-rays. Braggs (both father and son, received the Nobel Prize on 1915 for this) derived a simple equation relating the wavelength of *X*-rays to the angular positions of scattered beams. A crystal may be regarded as a stack of parallel planes of atoms. The atomic planes are often called Bragg Planes. Every crystal has several sets of Bragg planes oriented in different directions. Let us consider a series of Bragg planes separated by a distance *d*. Let

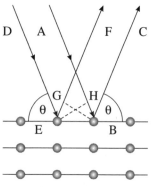

Fig. 28(a) Reflection of *X*-rays from crystal plane.

a beam of monochromatic *X*-rays represented by the parallel lines *AB* and *DE* be incident on these planes as shown in Figure 28(a). The scattered beams emerge along *BC* and *EF*. Here we are considering only the contributions of two adjacent atoms in the same plane. The rays *BC* and *EF* are coherent and reinforce each other if they are in phase. It requires that the path lengths *BH* and *EG* are equal, which will happen only when (as shown in Figure 28(a)). This is the condition of reflection. Now, the contributions of two adjacent atoms in successive planes, as shown in Figure 28(b) below, *MN* and *PQ* are considered.

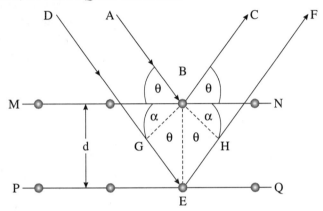

Fig. 28(b)   Reflection of X-rays from two consecutive lattice planes.

The path difference Δ between the rays reflected along *BC* and *EF* is, $\Delta = GE + EH$.

The rays *BC* and *EF* will be in phase only when,

$$\Delta = m\lambda \ (m = 0, 1, 2, 3, \dots)$$

Now, $\angle ABG = \theta + \alpha = 90°$ $(\because \angle ABG = \angle EGB = 90°)$ ...(1)

BE is normal to *MN*.

So, $\angle MBE = 90°$; $\alpha + \angle GBE = 90°$ ...(2)

From eqns. (1) and (2),

$$\angle GBE = \theta \qquad\qquad\qquad ...(3)$$

Also, from $\triangle BGE$,

$$\sin \theta = \frac{GE}{BE} = \frac{GE}{d} \ ,$$

$\Rightarrow$ $Ge = d \sin \theta$ ...(4)

Similarly,

$$EH = d \sin \theta \qquad\qquad\qquad ...(5)$$

Hence, using eqns. (4) and (5), the path difference is

$$\Delta = GE + EH = 2 \ d \sin \theta$$

Thereby, for the rays in phase,

$$\boxed{2 \ d \sin \theta = m\lambda} \qquad\qquad\qquad ...(6)$$

The above equation (6) for the condition for reinforcement of scattered wave is known as **Bragg Equation or Bragg's Law**. Here, θ represents the angle between the **Bragg plane** and the incident beam, also known as the **Glancing angle or Bragg angle**. 2θ is the angle of Diffraction and m is the order of reflection.

II. Laue Method

Laue, for the first time, predicted that a crystal acts as a natural three dimensional grating for X-rays where the regularly placed periodic lines of atoms serve as a function of parallel rules lines. He received the Nobel on 1914. Continuous X-rays produced by an X-ray tube are defined into a narrow beam by a set of lead screens S_1 and S_2 having pinholes at their centers as in Figure 29 below. A thin crystal C is mounted in the path of

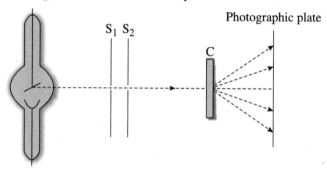

Fig. 29   Schematic diagram of Laue method assembly.

X-ray and a photographic plate is positioned beyond it. The single crystal, C, is placed at proper orientation relative to incident X-ray beams. As the X-rays penetrate C, some of the rays are scattered or deflected away by atoms from their initial direction. The scattered X-rays emerge from the crystal in specific directions as highly narrow beams and they are intercepted by the photographic film. On developing the film a pattern of bright spots corresponding to maxima are observed. They are shown in Figure 30 below. The pattern of Laue spots is a unique characteristic of the crystal C. The central bright spot on the film corresponds to the main unscattered beam. This method has been subsequently used to determine the crystal structures

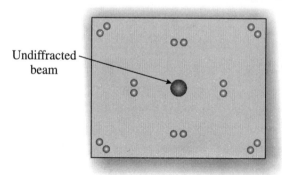

Fig. 30   Photographic plate showing Laue spots.

using X-rays of known wavelength. Each spot in the Laue diffraction pattern corresponds to interference maxima for a set of crystal planes satisfying the Bragg's equation for a particular wavelength selected from the continuous incident beam. From the positions and intensities of these spots, one can determine crystal structure, *i.e.*, the value of primitives, a, b, c and the details of unit cell of the solid concerned.

III. Powder Crystal Method

Investigations of crystal structures by using the Laue's method, is possible only when the material is available in the form of single crystals of reasonable size. However, for a large number of materials it is impossible to obtain a single crystal of required size. For such materials powder photography is highly suitable. It is also known as Debye-Scherrer method. The Figure 31 next shows the principle of the Powder method.

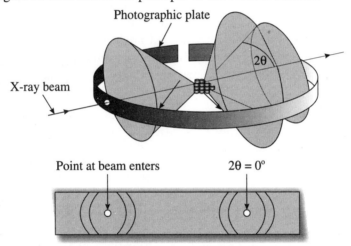

Fig. 31 Schematic diagram of powder method assembly along with diffraction spots on photographic plate.

Here, the material is crushed into a fine grain powder and compressed into a thin rod or sometimes packed into a small capillary tube. This is shown in Figure 31. A strip of photo film wrapped in opaque paper is mounted round the inside of a cylindrical drum. The specimen is positioned vertically at the centre of the drum. A narrow beam of monochromatic X-rays enters and leaves the drum through the apertures on opposite sides of the drum. The powder contains millions of tiny crystals oriented at random in all possible directions. Each crystallite has the same system of atomic planes. So, some of them are bound to lie with their planes at a glancing angle θ to the incident ray such that, Bragg's equation is satisfied. Each such a crystal will produce a spot on the photographic plate. Reflections will be produced by all such crystallites whose normal to the plane form a cone as shown in Figure above. Consequently the reflected rays will lie in the cone of semi-angle 2θ. The intersection of the cone formed by the reflected rays with the photographic plate yields a circle. On the film strip we thus see arcs. After exposure the various possible values of 2θ are calculated from the positions of the arcs and the radius of the camera drum.

If 'x' be the distance at which a reflected beam strike from the centre 'O' and 'R' be the radius of the drum, then,

$$2\theta = \frac{180}{\pi}\frac{x}{R} \text{ deg}$$

$$\left[2\theta \approx \tan 2\theta = \frac{x}{R} \Rightarrow 2\theta = \frac{x}{R} \text{ Radian} = \frac{x}{R}\frac{180}{\pi} \text{ degree}\right]$$

(as seen in Figure 32). $\Rightarrow \theta = \dfrac{90x}{\pi R}$ degree.

Using this value of θ into the Bragg's equation, 'd' can be calculated. This method helps to differentiate between amorphous to crystalline materials. Amorphous do not have any reflecting planes. Therefore, diffraction rings are not produced on the film. However, they may produce a smeared ring as they have a very short range periodicity of atoms.

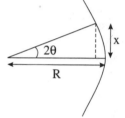

Fig. 32 Diagram of find diffraction angle.

4. ▍ DEFECTS IN SOLIDS

An ideal crystal does have a very long range of order in their atomic arrangement. However, ideal crystal neither occurs in nature nor can be produced by artificial methods. The deviations from an ideal crystalline structure are called defects. Defects can be classified into the following categories :

(a) Point defects (zero dimensional defects),

(b) Line defects (one dimensional defects), *e.g.*, dislocations,

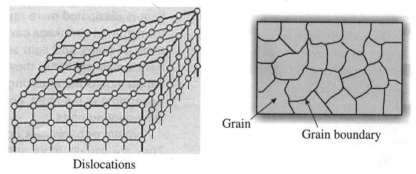

Dislocations

Grain

Grain boundary

Fig. 33 Figures shown dislocation of lattice plane and grain boundary.

(c) Surface defect (two dimensional defects), *e.g.*, grain boundary, and

(d) Volume defects (three dimensional defects), *e.g.*, precipitates.

I. Point Defects

It can be classified into three categories. However, these all three point defects does not change the stoichiometry of the crystal.

1. Vacancies : A vacancy is an atomic site from which the atom is missing as shown in Figure 34.

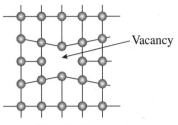

Fig. 34 Vacancy in a crystal lattice.

Vacancies are produced during crystallization as a result of thermal stress of atoms at high temperature. The atom then leaves the crystal and produces a vacancy in the lattice. This defect is known as vacancy defect. These vacancies often arise in the metallic structures (closed pack). In ionic crystals the formation of vacancy requires a local adjustment of charge such that charge neutrality is maintained in the crystal. Therefore a pair of vacancies arise which lead to missing of one cation and one anion from the structure. Such pairs of vacant sites are called a Schottkey Defect. The atoms surrounding a vacancy produce a tensile stress as evident from the above Figure.

2. Interstitials : An interstitial defect is an atom or an ion that has moved from its actual position to a place between angular lattice sites. The interstitial may be either a normal atom or a foreign atom.

When a normal atom moves into interstices it leaves behind its original position which is now vacant and thus produces a pair of defects. In ionic crystals, a cation goes into an interstitial position. Thus creates a cation - anion pair. This is known as **Schottkey Defect** as shown in Figure 35(a). A foreign atom can go into an interstitial space only when it is substantially smaller that the host atom. Therefore normally anion leavs its parent site and occupy intersitial space. These vacancy (created by leaving) - interstitial pair is called a **Frenkel Defect** as shown in Figure 35(b).

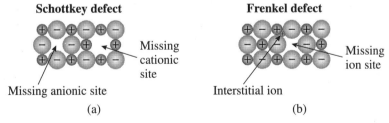

Schottkey defect — Missing cationic site — Missing anionic site — (a)

Frenkel defect — Missing ion site — Interstitial ion — (b)

Fig. 35   (a) Schottkey defect (b) Frenkel defect.

3. Impurity Defects : Impurity defects are foreign atoms introduced into the crystal lattice, either as an interstitial atom or a substitutional atom. If a host lattice is replaced by a foreign atom then a substitutional defect is observed. The substitutional atom replaces the host atom from its position. A **small** substitutional atom causes a tensile stress in a lattice as in Figure 36(a) below.

A large substitutional atom causes a compressive stress in a lattice as in Figure 36(b) below.

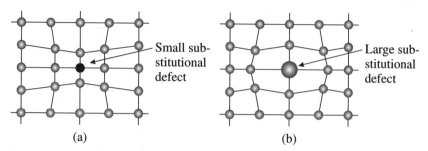

Fig. 36 Crystal defects (a) Small substitutional defect (b) Large substitutional defect.

4. Electronic Defect : This defect arises due to excess of electrons or deficiency of electrons in solid, which creates errors in charge distribution in solid. This means solid departure from its neutral charge distribution due to the electronic defects. For example when phosphorous is diffused in silicon, it occupy silicon site. Since it has five valence shell, four out of them form covalent bond with neighbour electrons of silicon atom leaving behind one electron as free. This free electron disturbs the charge neutrality of solid. In this fashion thousand of atoms diffuse in solid, which create appreciable change in the charge distribution in solid. This type of semiconductor is called as n-type semiconductor. This defect is intentionally introduced in the formation of various semiconductor devices like p-n junction diode, p-n-p or n-p-n transistor, MOS devices, CMOS devices etc.

5. Colour Centre : This defect arises when colourless crystal is heated in the presence of alkali metals like sodium, potassium and then quenched. These metals get ionized and positive charged atom occupy normal cation site in crystal while electron is trapped in anion site. This electron is weakly bound to this site and lead to the absorption in the visible range. Such centres at which electrons are bound to anion vacant site are called as colour centre or F-centre. For example if sodium chloride crystal is heated in the presence of sodium or potassium vapours and then quenched. Colourless sodium chloride crystal starts to appear as yellow due to the presence of F-centre.

II. Schottkey Defect (Quantitative)

Here, a vacancy is created in a metal (ionic crystal) by the migration of a metal atom / ion to the surface. So in a binary ionic crystal (A^+B^-, a cation – anion pair) a cation – anion pair will be missing from the respective lattice positions. This creation of such sites is called the **Schottkey Defect**.

The energy absorbed during this creation is more than compensated by the resultant disorder in the lattice or structure. The vacancies are always present in a crystal due to thermal condition variations. This process requires energy. Moreover, as disorder increases entropy of the system also increases. In thermal equilibrium a certain number of lattice vacancies are always present. Let 'n' be the concentration of Schottkey Defects produced by receiving nA^+ and nB^- ions. Also we consider, N and $\overline{E}$ are the total concentrations of cation-anion pairs and average energy required to produce a single Schottkey Defect respectively. This $\overline{E}_s$ is the energy required to take an atom from a lattice site inside the crystal to a lattice site on the surface, also called the 'enthalpy' of formation of the point defect.

So, $\Delta E = n\overline{E}_s$, is the increase in energy associated with n vacant sites.

The total number of different ways in which n Schottkey Defects can be produced is

$$W = N_{C_n} \times N_{C_n}$$

$$\Rightarrow \qquad W = \left[\frac{N!}{(N-n)! \, (n!)}\right]^2$$

The exponent 2 is for both cations and anions, as each can be removed in $\left[\dfrac{N!}{(N-n)! \, (n!)}\right]$ ways.

Thus, the increase in entropy is given by

$\Delta S = k \log W$, where k is Boltzmann's constant.

Hence, $\qquad \Delta S = k \log \left[\dfrac{N!}{(N-n)! \, (n!)}\right]^2$

$$\Rightarrow \qquad \Delta S = 2k \log \left[\frac{N!}{(N-n)! \, (n!)}\right]$$

This produce a change in Helmholtz free energy, H, So,

$$H = \Delta E - T\Delta S$$

$$\Rightarrow \qquad H = n\overline{E}_s - 2\,kT \log \left[\frac{N!}{(N-n)! \, (n!)}\right] \qquad \ldots(1)$$

Using Stirling's approximation (as $\log x! \cong x \log x - x$) we have from equation (1),

$$H = n\overline{E}_s - 2\,kT\,[\log (N!) - \log (N-n)! - \log (n!)]$$

$$\Rightarrow \qquad H = n\overline{E}_s - 2\,kT\,[(N \log N - N) -$$

$$\{(N-n) \log (N-n) - (N-n)\} - (n \log n - n)]$$

$$\Rightarrow \qquad H = n\overline{E}_s - 2\,kT\,[N \log N - (N-n) \log (N-n) - n \log n] \quad \ldots(2)$$

At equilibrium, the Free energy is constant, so,

$$\left.\frac{\partial H}{\partial n}\right|_T = 0$$

$$\Rightarrow \qquad \overline{E}_s = 2\,kT\left[0 - \left\{(-1) \log (N-n) + \frac{(-1)\,(N-n)}{(N-n)}\right\} - \log n - 1\right]$$

$$\Rightarrow \qquad \overline{E}_s = 2\,kT\,[\log (N-n) + 1 - \log n - 1]$$

$$\Rightarrow \qquad \overline{E}_s = 2\,kT \log \left(\frac{N-n}{n}\right)$$

$$\Rightarrow \qquad \log \left(\frac{N-n}{n}\right) = \frac{\overline{E}_s}{2\,kT}$$

$$\Rightarrow \qquad \frac{N-n}{n} = e^{\overline{E}_s / 2\,kT}$$

When, $n \ll N$, $(N - n) \approx N$, hence,

$$N = n e^{\overline{E}_s / 2kT} \quad \Rightarrow \quad n = N e^{\overline{E}_s / 2kT}$$

Hence, the number of Schottkey defects, n, depends on

(i) total number of ion-pairs, N,

(ii) average energy required to produce a Schottkey defect pair, $\overline{E}_s$, and

(iii) temperature of the material.

III. Frenkel Defect (Quantitative)

Here, a vacancy is created in any material by the migration of an atom/ion to the surface. So in a material vacancy – interstitial pair is formed. This creation of such pairs is called the **Frenkel Defect**.

The energy absorbed during this creation is more than compensated by the resultant disorder in the lattice or structure. The vacancies are always present in a crystal due to thermal condition variations. This process requires energy. Moreover, as disorder increases entropy of the system also increases. In thermal equilibrium a certain number of lattice vacancies are always present. Let 'n' and N are the concentrations of Frenkel Defects produced and the total number of cations present in the material respectively. Also we

consider, N_i and $\overline{E}_f$ are the total concentration of interstitial sites and average energy required to produce a single Frenkel Defect respectively.

The total number of different ways in which n Frenkel Defects can be produced is

$$W = \left[\frac{N!}{(N-n)! \, (n!)} \right] \left[\frac{N_i!}{(N_i - n)! \, (n!)} \right]$$

The first part on RHS represents the no. of ways the n cation or anion vacancies are formed among total N ions and the second part gives the no. of ways the n vacancies can combined with N_i interstitial sites. Thus, the increase entropy is

$$\Delta S = k \log W$$

$$\Rightarrow \qquad \Delta S = k \log \left[\frac{N!}{(N-n)! \, (n!)} \right] \left[\frac{N_i!}{(N_i - n)! \, (n!)} \right]$$

This produce a change in Helmholtz free energy, H, as

$$H = \Delta E - T\Delta S$$

$$\Rightarrow \qquad H = n\overline{E}_f - kT \left[\log \frac{N!}{(N-n)! \, n!} + \log \frac{N_i!}{(N_i - n)! \, n!} \right] \quad \ldots(1)$$

Using Stirling's approximation we have from (1),

$$H = n\overline{E}_f - kT \, [\log (N!) - \log (N-n)! - \log (n!) + \log (N_i!)$$
$$- \log (N_i - n)! - \log (n!)]$$

$$\Rightarrow H = n\overline{E}_f - kT \, [(N \log N - N) - \{(N-n) \log (N-n) - (N-n)\}$$
$$- 2 \, (n \log n - n) + (N_i \log N_i - N_i) - ((N_i - n) \log (N_i - n) - (N_i - n)]$$

$$\Rightarrow H = n\overline{E}_f - 2 \, kT \, [N \log N + N_i \log N_i - (N-n) \log (N-n)$$
$$- (N_i - n) \log (N_i - n) - 2n \log n]$$

At equilibrium, the Free energy is constant with respect to defect concentration, so,

$$\frac{\partial H}{\partial n}\bigg|_T = 0$$

$$\Rightarrow \overline{E}_f = kT\left[0 + 0 - \left\{(-1)\log(N-n) + \frac{(-1)(N-n)}{(N-n)}\right\}\right.$$

$$\left. - \left\{(-1)\log(N_i-n) + \frac{(-1)(N_i-n)}{(N_i-n)}\right\} - 2(\log n + 1)\right]$$

$$= kT\left[\log(N-n) + 1 + \log(N_i-n) + 1 - 2\log n - 2\right]$$

$$= kT\log\left(\frac{(N-n)(N_i-n)}{n^2}\right)$$

When, $n \ll N, (N-n) \approx N$ and $n \ll N_i$,

then, $\Rightarrow$ $\overline{E}_f = kT\log(N_i N) - 2\,kT\log n$

$\Rightarrow$ $\log n = kT\left(\dfrac{\log(N_i N)}{2\,kT}\right) - \dfrac{\overline{E}_f}{2\,kT}$

$\Rightarrow$ $\log n = \log(\sqrt{N_i N}) - \dfrac{\overline{E}_f}{2\,kT}$

$\Rightarrow$ $n = (N\,N_i)^{1/2}\,e^{-\overline{E}_f/2\,kT}$

Hence, the number of Frenkel defects, n, depends on

(i)   proportional to $\sqrt{NN_i}$,

(ii) average energy required to produce a Frenkel defect pair, $\overline{E}_f$, and

(iii) temperature of the material.

5. ELEMENTARY IDEAS OF QUARKS AND GLUONS

Quarks

In 1803 British scientist John Dalton showed that any materials are chemical compounds formed by mixing units in fixed proportion. He named this minimum unit as 'Molecule'. Later on it was found that Atoms are joined together to form molecule. However, from the very beginning there was a doubt regarding the indivisibleness of atom. J.J. Thompson of Cambridge proved the existence of Electron in atom by using thermionic emission. In 1911 E. Rutherford first labeled the atomic structure indicating that an atom consists of a positively charged 'Nucleus' and negatively charged 'Electrons'. The electrons are revolving around the nucleus in their own trajectory.

Initially it was thought that nucleus only consisted of positively charged Protons. Then, in 1932, colleague of Prof. Rutherford, James Chadwick discovered 'Neutrons', the chargeless particle residing with proton in the nucleus of an atom. Up to 1968 it was thought that these protons and neutrons along with electrons are the only elementary particles. But with the development of technology for yielding large amount of energy for high energy research it was possible to make collisions between proton and also between protons and neutrons. These experiments revealed the elementary detail of proton and neutrons. The physicist of Caltech M. Gellman received Nobel Prize for this invention and he named the particles as 'Quarks'. There are six types of quark available till date, namely, Top, Bottom, Up, Down, Strange and Charmed. Each quark also have three colours, *e.g.*, Red, Green and Blue. A proton consists of two up and one down quark, a neutron consists of one up and two down quarks, and so on. Other particles can also be made of other quarks, but the mass of them are very large and they decay quickly to form proton and neutron. However, electron remains the elementary particle. Gellmann and Zivag proposed that all strongly interacting particles are built up of three new particles, called **quarks.** They are assumed to have fractional charges, fractional baryon numbers and spin of 1/2 each. As they have spin 1/2 they all are Fermions.

Originally three quarks were proposed (i) UP quark (u), (ii) DOWN quark (d) and (iii) STRANGE quark (s).

Another three quarks were then added in their family later on. These are (iv) CHARM © (v) BEAUTY (or BOTTOM) (b) and (vi) TRUTH (or TOP) (t) quarks.

Properties :

Name	Charge	Isospin		Baryon number	Strangeness	Charm	Bottom	Top
		T_1	T_2					
d	–1/3	1/2	–	1/3	0	0	0	0
u	+2/3	1/2	1/2	1/3	0	0	0	0
s	–1/3	0	1/2	1/3	–1	0	0	0
c	+2/3	0	0	1/3	0	1	0	0
b	–1/3	0	0	1/3	0	0	–1	0
t	+2/3	0	0	1/3	0	0	1	1

Gluons

Politizer and Wilezek discovered (1974) that in many circumstances the effective strength of interaction became weaker at short distances. This interaction between the quarks becomes weaker within short distances. According to this theory the quarks of a given type create a field (called the **Gluon field**) around them so that they emit or reabsorb a type of hypothetical particles, much like the emission and re-absorption of virtual protons by electrically charged particles. These hypothetical particles are called **Gluons.**

Properties :

1. Isospin = 0
2. Other properties resemble with that of protons, *i.e.*, charge, magnetic moment and charge parity (= –1)
3. Unlike electrically neutral particles, the Gluons carry a type of charge, called the COLOUR.

The Gluon Field is non-linear and increases with increasing distance from the quark. The effective COLOUR charge of quark also increases in the process. The interaction between quark disappear at short distance.

2

QUANTUM PHYSICS

The most flashing phenomenon of science developed in the twentieth century is the birth of quantum physics. Quantum physics completely revolutionized the realization or the thought processes related to every phenomenon especially in the atomic scale. It generates dramatic changes about scientific approaches towards the understanding of almost anything in the universe. The atomic behaviour is not an ordinary thing like what we are observing in day to day life as our human intuition become normally applicable to the larger objects. For an example, the yellowish colour of gold is its characteristic colour. As we are going to reduce the size of gold particles we will find that there is no actual existence of the colour of gold or any material. As we reduce the size of gold particles to 30-500 nanometer, it gives metallic colour of crimson to blue, in the range of 3-30 nm, it turns into red metallic and transparent one. Less than 3 nm range it gives non-metallic orange colour and in atomic dimension of 1 A we can only get colourless atoms. Hence in the reduction up to the nanoscale atomic dimension we cannot explain the phenomena by using our old or classical physics. Only quantum physics can be able to explain these phenomena. If we can say that we have two openings in front of a gun and we can fire only one bullet from the gun, we expect that the bullet can either pass through any one of them or may not pass at all. However, as per quantum physics in its domain there is fine possibility or probability of the bullet to pass through both the openings at the same time. It sounds absolutely rubbish to us as we are all living in a classical world. Even this probability based concept of quantum physics cannot satisfy even Einstein till the end of his life. In his famous statement "God does not ply dice with the universe" he illustrated his utter hopelessness regarding quantum physics by the element of 'Chance' involved in it although he is the pioneer of it. The successes of the application of it in experiments after experiments (even after death of the prodigy) disturbed him very much to force him to make such a comment. However, most coveted physicist of modern time, Stephen Hawking, fittingly commented "Einstein was confused, not the quantum theory". The slow but steady progress about the knowledge about the atomic behaviour during first quarter of twentieth century finally resolved in 1926 and 1927 by Schrodinger, Heisenberg and Born. It helped to create a conceptually concrete description about nanoscale materials. As we all now expect the twenty first century as the century of nanomaterials, we should have a complete understanding of quantum physics as a building block.

1. DIFFICULTIES WITH CLASSICAL PHYSICS

Classical mechanics is based on Newton's three laws. It can explain correctly the motion of planets, stars and macroscopic along with microscopic terrestrial bodies moving with non-relativistic speeds (particle velocity, $v \ll c$, velocity of light). It is able to explain successfully the motion of objects which are directly observable or exist in macroscopic world. But classical concepts can not be applied in the region of atomic dimension or in microscopic world. There the dynamic variable like energy and momentum did not have the same meaning as in classical dynamics. These variables were found to have a discrete value in a particular state of atom and did not change in continuous manner as in the case of classical laws. These new concepts led to a new theory called Quantum Mechanics. The adequancies of classical theory are :

I.   It does not hold in the region of atomic dimension, *i.e.*, the relativistic motion of atoms, electrons, protons, etc.

II. According to Rutherford, atom consists of positively charged nucleus of very small dimension and it is surrounded by negatively charged electrons of much smaller mass. These electrons revolve in circular orbits around the nucleus. We know any moving charged particle always radiates energy. So during revolution electron will loose energy in the form of electromagnetic wave and ultimately by orbiting in spiral paths they will collapse in nucleus. This leads to the instability of atom in contradiction with the fact of the stability of atoms.

III. It could not explain observed spectrum of blackbody radiations. Lumer and Pringsheim investigated experimentally the distribution of energy among the radiation emitted by a blackbody at different temperatures.

Experimental observations :

(i) The energy is not uniformly distributed in radiation spectrum of a blackbody as shown in Fig. 1.

(ii) At a given temperature (T) the intensity of radiation (E_λ) increases with λ and at a particular λ (*i.e.*, λ_m), E_λ is maximum and then decreases.

(iii) With increase in T, λ_m decreases. So,

$$\boxed{\lambda_m T = \text{Constant}}$$

This is **Wien's Displacement Law.**

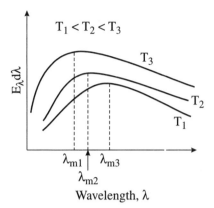

Fig. 1 Experimentally observed energy distribution curved at different temperatures.

(iv) With increase of T, E_λ increases for all λ.

(v) Total energy (E) emitted by the body at a particular T is given by the area of the curve.

So, $\boxed{E \, \alpha \, T^4}$: This is **Stefan's Law**.

Rayleigh and Jeans (R-J) derived an expression for the radiation energy per unit volume of a very hot body by using Classical theory. According to R-J the energy emitted by a blackbody increases with ν and becomes infinite at large ν. The experimentally observed curve however shows that the energy radiated at a particular T becomes zero both at low and high ν as shown in Fig. 2. Also, R-J Law indicates that the total energy

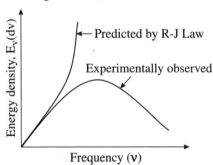

Fig. 2 Experimentally observed and theoretical R-J's energy distribution curves.

emitted by a blackbody per unit area per unit time is proportional to ν at all T. These are all totally FALSE conclusions.

Based on the classical electromagnetic theory Wein also obtained the expressions for energy density in terms of wavelength. This was found to be valid only in low wavelength or high frequency region, while yielded wrong results in other wavelength region. All these discrepancies was later on removed by Max Planck on the basis of Quantum Theory.

(IV) Dulong and Petit calculated the specific heat of metal using the classical theory. Only at high temperature it is found that specific heat is independent of temperature. However at low temperature it failed to explain the specific heat variation for metal and gases.

(V) It could not explain the origin of discrete spectra of atoms. The excited atom of hydrogen emits the electromagnetic radiation of certain definite wavelength and not all the wavelength continuously emitted as expected from classical theory.

(VI) It could not explain a large no. of phenomena like Photoelectric Effect, Compton Effect, etc.

All these inadequacies of classical theory led to the development of quantum mechanics.

2. PLANCK'S RADIATION LAW AND DISCOVERY OF PLANCK'S CONSTANT

Planck derived an empirical formula to explain the observations on the basis of following assumptions known as **Planck's Hypothesis.**

(i) A blackbody radiation chamber is filled up with radiations along with **Simple Harmonic Oscillators (SHO)** of molecular dimensions (called as Planck's Oscillators or Resonators) which can vibrate with all possible frequencies.

(ii) The resonators cannot radiate or absorb energy continuously, but an resonator of frequency v **can radiate or absorb energy in quanta of magnitude** hv; where, h is a universal constant, known as **Planck's constant** $(h = 6.62 \times 10^{-34}$ **J-sec**$)$. It implies the exchange of energy between radiation and matter is limited to discrete set of values 0. $hv, 2hv, \ldots, nhv$; where n is an integer. This energy is a multiple of some small unit hv, called the "Quantum".

Derivation of Planck's Radiation Law

First Step : To derive first the nos. of modes of vibrations per unit volume lying in the frequency range v and $v + dv$.

Let the radiation of wavelength λ enclosed between two perfectly reflecting parallel walls separated by a distance l from stationary waves, then,

$l = n_1 \times \dfrac{\lambda}{2}$; n_1 is the nos. of nodal planes between parallel walls along one axis as shown in Fig. 3(a). So, frequency, v along any one axis $i.e.$, X-axis,

is $v = \dfrac{c}{\lambda} = \dfrac{c}{2l/n_1} = \dfrac{cn_1}{2l}$. Similarly, frequencies v along any Y & Z-axis, are

$\dfrac{cn_2}{2l}, \dfrac{cn_3}{2l}$. The total nos. of n's are the nos. of points lying in one octant between the spheres of radii v and $v + dv$ as shown in Fig. 3(b). Volume of this spherical shell $= 4\pi v^2 dv$. The volume of an elementary cubic lattice corresponding to one vibration is $\left[\dfrac{cn_1}{2l} \times \dfrac{cn_2}{2l} \times \dfrac{cn_3}{2l}\right]$ [with $n_1 = n_2 = n_3 = 1$]

$= \left(\dfrac{c}{2l}\right)^3$. So, the total no. of independent vibrations lying between frequency v and $v + dv$ in the cube of volume l^3 is $8 \cdot \left(\dfrac{4\pi v^2 \, dv}{c^3}\right)(l)^3$.

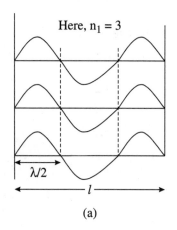

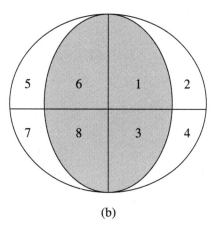

(a)　　　　　　　　　　(b)

Fig. 3

Hence, no. of independent vibrations per unit volume is given by

$$= \frac{4 \pi v^2 \, dv}{c^3}$$

[Division by 8 is there as n_1, n_2 and n_3 occur in one out of 8 octants]

The blackbody radiation travel with velocity of light c and are transverse in character. The no. of modes of vibration of transverse waves is double as for longitudinal wave. So for a blackbody radiation the nos. of modes of vibration of transverse waves is double as for longitudinal wave. For a blackbody radiation the no. of modes of vibration per unit volume in frequency range v & $v + dv$ is

$$2 \times \frac{4\pi v^2 \, dv}{c^3} = \frac{4\pi v^2 \, dv}{c^3} \qquad \text{...(1a)}$$

Second Step : To find the average energy of a Planck's Oscillator

According to Planck's hypothesis of Quantum theory the average energy of Planck's Oscillator is ϕ.

Let N be the total no. of Planck's Oscillator and E their total energy, then average energy per Planck's Oscillator is

$$\phi = \frac{E}{N}.$$

Here, we consider $\varepsilon = h\nu$. According to Maxwell's Law of Molecular motion, if ε is a certain amount of energy, the probability that a system will have energies 0, ε, 2ε, ..., $r\varepsilon$, ... are in the ratio of $1 : e^{-\varepsilon/kT} : e^{-2\varepsilon/kT} : e^{-3\varepsilon/kT} : \dots$ etc.

Here, k is Boltzmann's constant and T is the absolute Temperature of the body.

If N_0 be the no. of resonators having energy zero, then the no. of resonators N_1 having an energy ε is $N_0 e^{-\varepsilon/kT}$; the no. of resonators N_2 having an energy 2ε is $N_0 e^{-2\varepsilon/kT}$; and so on.

Hence, $\qquad\qquad N = N_0 + N_1 + N_2 + \dots$

$\Rightarrow \qquad\qquad N = N_0 + N_0\, e^{-\varepsilon/kT} + N_0 e^{-2\varepsilon/kT} + \dots$

$\Rightarrow \qquad\qquad N = N_0 [1 + y + y^2 + \dots] = \dfrac{N_0}{1 - y}$

$\qquad\qquad$ (using $y = e^{-\varepsilon/kT}$ & GP sum formula for $y < 1$)   ...(1)

Now, total energy of Planck's Oscillator will be

$$E = N_0 \times 0 + N_1 \times \varepsilon + N_2 \times 2\varepsilon + \dots$$

$\Rightarrow \qquad\qquad E = N_0\, \varepsilon\, y + e\, N_0\, \varepsilon\, y^2 + 3N_0\, \varepsilon\, y^3 + \dots$

$\Rightarrow \qquad\qquad E = N_0\, \varepsilon\, [y + 2y^2 + 3y^3 + \dots] \qquad\qquad$ ...(2)

Let, $\qquad\qquad S = [y + 2y^2 + 3y^3 + \dots] \qquad\qquad$ ...(3)

Hence, $\qquad Sy = [y^2 + 2y^3 + 3y^4 + \dots] \qquad\qquad$ ...(4)

Using eqns. (3) & (4),

$$S - Sy = [y + y^2 + y^3 + y^4 + \dots]$$

$\Rightarrow \qquad\qquad S(1 - y) = y[1 + y^2 + y^3 + y^4 + \dots] = \dfrac{y}{1 - y}.$

$\Rightarrow \qquad\qquad S = \dfrac{y}{[1 - y]^2};$

Hence, from eqn. (2)

$$E = \varepsilon S N_0 = \varepsilon N_0 \frac{y}{[1 - y]^2}.$$

So, the average energy per Planck's Oscillator is

$$\phi = \frac{E}{N} = \frac{\varepsilon N_0 \left[\dfrac{y}{(1 - y)^2} \right]}{N_0 / (1 - y)}$$

$$\Rightarrow \qquad \phi = \varepsilon \frac{y}{1-y}$$

$$\Rightarrow \qquad \phi = \frac{\varepsilon\, e^{-\varepsilon/kT}}{1 - e^{-\varepsilon/kT}}$$

$$\Rightarrow \qquad \boxed{\phi = \frac{\varepsilon}{e^{\varepsilon/kT} - 1}} \qquad \ldots(5)$$

The energy density belonging to the frequency range dv can then be obtained as $E_v\, dv = \phi \times$ no. of modes of vibrations per unit volume (Using eqns. (1A) and (5))

$$\Rightarrow \qquad \boxed{E_v\, dv = \frac{8\pi v^2\, dv}{c^3} \times \frac{\varepsilon}{e^{\varepsilon/kT} - 1}}$$

$$\Rightarrow \qquad E_v\, dv = \frac{8\pi v^2\, dv}{c^3} \times \frac{hv}{e^{hv/kT} - 1} \qquad \ldots(6)$$

This is the **Planck's radiation Law** in terms of **frequency**.

For wavelength :

As $\qquad\qquad v = c/\lambda$

$$\Rightarrow \qquad dv = -\left(\frac{c}{\lambda^2}\right) d\lambda \approx \left(\frac{c}{\lambda^2}\right) d\lambda$$

[as we are only considering the magnitude]

Using eq. (6),

$$E_\lambda\, d\lambda = \frac{8\pi h}{c^3} \left(\frac{c}{\lambda}\right)^3 \frac{(c/\lambda^2)\, d\lambda}{e^{hc/\lambda kT} - 1}$$

$$\Rightarrow \qquad E_\lambda\, d\lambda = \frac{8\pi\, hc}{\lambda^5} \left[\frac{1}{e^{hc/\lambda kT} - 1}\right] d\lambda \qquad \ldots(7)$$

This is the **Planck's radiation Law** in terms of **wavelength**.

From Planck's Law

(a) Derivation of Wien's Law : For short wavelength, $e^{hc/\lambda kT} > 1$.

From eqn. (7)

$$\Rightarrow \qquad \boxed{E_\lambda\, d\lambda = \frac{8\pi\, hc}{\lambda^5} e^{-hc/\lambda kT}\, d\lambda} \quad \Rightarrow \quad \text{Wien's Law.}$$

(b) Derivation of R-J's Law : For long wavelength,

$$e^{hc/\lambda kT} \approx 1 + \frac{hc}{\lambda kT}$$

From eqn. (7)

$$\Rightarrow \qquad E_\lambda\, d\lambda \approx \frac{8\pi\, hc}{\lambda^5} \frac{d\lambda}{\left(1 + \dfrac{hc}{\lambda kT}\right) - 1}$$

$$\Rightarrow \qquad \boxed{E_\lambda \, d\lambda = \frac{8\pi \, kT}{\lambda^4} \, d\lambda} \quad \text{R-J's Law}$$

Verification of Planck's Law (Experimental)

In Fig. below we observe the energy distribution curve for wavelengths. Fig. 4 shows that Planck's curve is the accurate curve with a total agreement with the experimental curve.

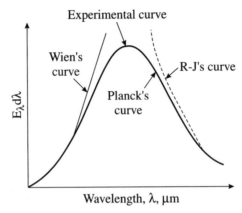

Fig. 4 Energy distribution curves obtained experimentally, using Planck's, Wien's and R-J's Laws.

Wien's law holds good for $\lambda < 2.2$ μm and Rayleigh-Jean's law holds good for larger wavelengths only.

Also, putting $\qquad x = \dfrac{h\nu}{kT} = \dfrac{hc}{\lambda \, kT}$,

we have,

R-J : $E_\lambda \, d\lambda = \dfrac{8\pi \, kT}{\lambda^4} \, d\lambda;$

Wien : $E_\lambda \, d\lambda = \dfrac{8\pi \, kT}{\lambda^4} \dfrac{x}{e^x} \, d\lambda;$

Planck : $E_\lambda \, d\lambda = \dfrac{8\pi \, kT}{\lambda^4} \dfrac{x}{e^x - 1} \, d\lambda.$

3. | QUANTUM THEORY : SIMPLE CONCEPTS

Planck's quantum hypothesis led to the conclusion that radiation is emitted in discrete packets, not in continuous fashion. These packets of energy are called 'quanta' of photons. So, photon nature of radiation regards that radiation exhibits a wave phenomena, but its energy contents are given to atoms in form of quanta. If the photon is regarded as a particle of radiation then we have to find out its energy, mass, etc.

I. Energy

Energy of photons is only in multiples of $h\nu$. Here h is Planck's constant and ν is its frequency. When photons undergo interaction with matter they completely or partially transfer their energies and their frequencies are thus lowered, thus maintaining particle character. So the intensity has nothing to do with photon energies, it simply gives their number. More photons mean more energy. The energy is dependent on the intrinsic properties of photons, *i.e.*, frequency. This is **contrary to the classical concept** when the radiation is considered as wave and energy is estimated by the intensity of wave.

Planck's constant (h) : This is the elementary quantum, responsible for discreteness of the photon to behave as a particle.

$$h = \frac{E}{\nu} \text{ ; dimension of}$$

$$h = \frac{MLT^{-2} \times L}{T^{-1}} = M\,[LT^{-1}]\,L = mvr \text{ (angular momentum)}.$$

Hence 'h' defined as the smallest quantum of angular momentum of a particle. Further Bohr assumed that electrons of an atom could revolve only those orbits whose angular momentum is an integral multiple of $h/2\pi$ (*i.e.*, h).

II. Mass and Momentum of Photon

Using the Theory of Relativity, $E = mc^2$, c is the velocity of light in free space.

$$\Rightarrow \qquad m = \frac{E}{c^2} = \frac{h\nu}{c^2}.$$

So, momentum $= $ mass $\times$ velocity $= \dfrac{h\nu}{c}$.

$$\text{Energy} = h\nu = mc^2 = \frac{m_0}{\sqrt{1 - \dfrac{v^2}{c^2}}} c^2$$

$$\Rightarrow \qquad m_0 = \left[\sqrt{1 - \frac{v^2}{c^2}}\right]\left(\frac{h\nu}{c^2}\right).$$

Here m_0 and v are the rest mass and velocity of the photon respectively. Now, as photon travels with a velocity c, then $m_0 \to 0$. Hence photon is not a particle in complete sense, as all material particles must have rest mass. So photon has also the wave nature since wave does not have any mass, sometime it behaves in compact form like a particle.

III. Wave-Particle Duality

Planck's hypothesis gives satisfactory explanation of black body radiation, which suggested the quantized nature of electro magnetic radiations. Radiations also show particle like characteristic in showing Photoelectric effect, Compton effect, etc. This gives the evidence of their dual nature.

(A) Photoelectric Effect : This is the phenomenon of ejection of electrons from a metal surface when light of suitable wavelength falls on it. In this effect the phenomenon of no time lag between absorption of light and emission of electron, requirement of threshold frequency for photoelectric emission, dependence of kinetic energy of ejected electrons on frequency of incident radiation are not at all explained by the use of classical theory. In 1905 Einstein gave satisfactory explanation of photoelectric effect by using Planck's idea of energy quantization.

Features :

(i) Numbers of photoelectrons thus ejected is directly proportional to the intensity of incident light.

(ii) The velocity of photoelectrons increases with increase of the frequency of the incident light.

(iii) The light of frequency less than a certain frequency v_0 is not capable of ejecting electrons from metal surface.

According to Einstein's theory when an electromagnetic radiation of energy hv falls on the metal surface, its energy is used up in (a) ejecting the electron from the metal surface, ($\Rightarrow$ work function, ϕ_0), and (b) the rest of energy imparts to the kinetic energy (KE) to the ejected electrons.

Thus, $$E = hv = \phi_0 + \frac{1}{2} mv^2$$

But, $$\phi_0 = hv_0$$

So, $$hv = hv_0 + \frac{1}{2} mv^2$$

$\Rightarrow$ $$\boxed{\frac{1}{2} mv^2 = h(v - v_0)}$$

This is **Einstein's photoelectric equation**.

(B) Compton Effect : When high frequency or high energy radiations are scattered by the electrons of the scatterer the frequency of the scattered wave is smaller than the original radiations. This is Compton Effect. This change in frequency also suggests that the photons behaving as particles, not as waves as some of its energy is transferred after scattering (same way as a particle looses energy after collision).

4. DE-BROGLIE'S HYPOTHESIS : WAVE-PARTICLE DUALITY

At 1923, **De-Broglie** proposed that particles such as electron, proton, etc., might have dual nature as light (photon) does.

Hypothesis : Each material particle in motion behaves as waves and the wavelength λ associated with any moving particle of momentum p is given by

$$\boxed{\lambda = \frac{h}{p}}$$

So according to de-Broglie, $\lambda = \frac{h}{p} = \frac{h}{mv}$; where, m & v are mass and velocity of the particle.

If E be the energy of the wave of frequency v, so,

$$v = \frac{E}{h} = \frac{mc^2}{h} \quad \text{(as } E = mc^2, \text{ from Einstein's equation)}$$

the de-Broglie wave velocity u, then,

$$v = \frac{u}{\lambda} \implies u = v\lambda$$

$$\implies \qquad u = \frac{E}{h} \times \frac{h}{mv} = \frac{mc^2}{h} \times \frac{h}{mv} = \frac{c^2}{v} = c \times \frac{c}{v}.$$

Hence, as $\frac{c}{v} > 1$, then, $u > c$.

However, **this is not possible** as it means the de-Broglie wave associated with the particle travels faster than the particle itself.

This difficulty was later on resolved by Schrödinger by postulating that a material particle in motion is equivalent to a Wave Packet rather than a single wave.

5. WAVE PACKET : GROUP VELOCITY AND WAVE OR PHASE VELOCITY

A **wave packet** comprises of a group of waves, each with slightly different velocity and wavelength. With phase and amplitude so chosen that they interfere constructively over only a small region of space where the particle can be located, outside this region they interfere destructively to reduce their amplitude to zero rapidly as shown in Fig. 5. Such a packet moves with its own velocity and is known as the **Group velocity** (v_g). The individual wave forming the packet possesses an average velocity is known as the **Wave or Phase velocity** (u).

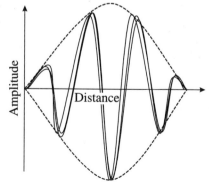

Fig. 5 A wave packet.

Hence, when a single wave of definite wavelength travels in a medium its velocity of propagation in the medium is only its Wave or Phase velocity. However, when a number of waves of different λ are moving with different velocities in a medium, the observed velocity is now the velocity of the wave packet (or wave group) formed by the waves. This is the Group velocity.

Theory : We now assume two wave trains have the same amplitude, but are slightly different frequencies ν & ν', and the phase velocities u & u'. So they are represented by

$$\psi_1(x, t) = A \sin w\left(t - \frac{x}{u}\right) = A \sin(\omega t - kx) \qquad \qquad ...(1)$$

$$[\text{as, } \omega \frac{x}{u} = \frac{2\pi}{\lambda} \times \frac{\lambda}{T} \times \frac{1}{u} \times x = k \times u \times \frac{1}{u} \times x = kx]$$

$$\psi_2(x, t) = A \sin \omega'\left(t - \frac{x}{u'}\right) = A \sin(\omega' t - k'x) \qquad \qquad ...(2)$$

where, ω & ω' are angular frequencies as $\omega = 2\pi\nu$ and $\omega' = 2\pi\nu'$. k & k' are the propagation constants given as,

$$k = \frac{2\pi}{\lambda} = \frac{2\pi}{T} \times \frac{T}{\lambda} = \frac{\omega}{\lambda/T} = \frac{\omega}{u};$$

Similarly, $$k' = \frac{\omega'}{u'}.$$

The superposition of these two wave trains gives a vibration $\psi(x, t)$, given by,

$$\psi(x, t) = \psi_1(x, t) + \psi_2(x, t)$$

$$\Rightarrow \psi(x, t) = A[\sin(\omega t - kx) + \sin(\omega' t - k'x)]$$

$$= 2A \sin\left[\frac{(\omega t - kx) + (\omega' t - k'x)}{2}\right]\cos\left[\frac{(\omega t - kx) - (\omega' t - k'x)}{2}\right]$$

$$\Rightarrow \psi(x, t) = 2A \cos\left[\frac{(\omega - \omega')}{2}t - \frac{(k - k')}{2}x\right]\sin\left[\frac{(\omega + \omega')}{2}t - \frac{(k + k')}{2}x\right]$$

$$...(3)$$

Equation (3) represents a vibration of amplitude $2A \cos\left[\frac{(\omega - \omega')}{2}\right.$ $\left. t - \frac{(k - k')}{2}x\right]$ angular frequency $\frac{(\omega + \omega')}{2}$ and propagation constant $\frac{(k + k')}{2}$.

So the wavelength, $$\boxed{\lambda = \frac{2\pi}{(k + k')/2} = \frac{4\pi}{k + k'}}$$

The phase velocity, $$\boxed{u = \frac{\omega}{k} = \frac{(\omega + \omega')/2}{(k + k')/2} = \frac{\omega + \omega'}{k + k'}}$$

The amplitude in eqn. (3) is $2A \cos\left[\dfrac{(\omega - \omega')}{2} t - \dfrac{(k - k')}{2} x\right]$ and it itself

is moving with velocity, v_g, along X-axis. Since this velocity of the amplitude is signal velocity, this is only the Group velocity.

When $\omega \to \omega'$, so, $k' \to k$; so,

$$u = \frac{2\omega}{2k} = \frac{\omega}{k} = \frac{2\pi v}{2\pi/\lambda} = v\lambda$$

and

$$v_g = \underset{\omega' \to \omega,\, k \to k'}{Lt} \frac{\omega - \omega'}{k - k'} = \frac{d\omega}{dk},$$

Hence,

$$u = \frac{\omega}{k} \quad \text{and} \quad v_g = \frac{d\omega}{dk} \qquad \qquad ...(4)$$

Group velocity equals the particle velocity : For a system where E and V are total and potential energy of the particle respectively, then,

$$\frac{1}{2} mv^2 = E - V \Rightarrow v = \sqrt{\frac{2(E - V)}{m}} \; .$$

Here, v and m are the velocity and mass of the particle respectively. Now, using de-Broglie formula,

$$\lambda = \frac{h}{mv}$$

$\Rightarrow$

$$\lambda = \frac{h}{m\sqrt{\dfrac{2(E - V)}{m}}}$$

$\Rightarrow$

$$\lambda = \frac{h}{m} \sqrt{\frac{m}{2(E - V)}} \qquad \qquad ...(5)$$

From eqn. (4),

$$v_g = \frac{d\omega}{dk} = \frac{d(2\pi v)}{d(2\pi/\lambda)} = \frac{dv}{d(1/\lambda)}$$

$\Rightarrow$

$$\frac{1}{v_g} = \frac{d(1/\lambda)}{dv}$$

$\Rightarrow$

$$\frac{1}{v_g} = \frac{d}{dv}\left[\frac{m}{h}\sqrt{\frac{2(E - V)}{m}}\right]$$

$\Rightarrow$

$$\frac{1}{v_g} = \frac{1}{h}\frac{d}{dv}\sqrt{2m(hv - V)}$$

$\Rightarrow$

$$\frac{1}{v_g} = \frac{1}{h}\frac{1}{2}[2m(hv - V)]^{-1/2}(2mh)$$

$\Rightarrow$

$$\frac{1}{v_g} = \frac{m}{\sqrt{2m(E - V)}} = \frac{1}{v}$$

$\Rightarrow$

$$\boxed{v_g = v}$$

Hence, the motion of a material particle is equivalent to a group of waves or wave packet.

General relationship (for non-relativistic particle) between v_g and u : From de-Broglie's hypothesis,

$$\lambda = \frac{h}{p} = \frac{h}{mv_g}$$

Also,

$$E = h\nu = \frac{1}{2} mv_g{}^2$$

$\Rightarrow$

$$\nu = \frac{mv_g{}^2}{2h}$$

Now,

$$u = \frac{\omega}{k} = \frac{2\pi\nu}{2\pi/\lambda}$$

$\Rightarrow$

$$u = \nu\lambda = \left(\frac{mv_g{}^2}{2h}\right) \times \left(\frac{h}{mv_g}\right)$$

$\Rightarrow$

$$\boxed{u = \frac{1}{2} v_g}$$

$\Rightarrow$ Phase velocity is half of the group velocity.

Now, for a dispersive medium, there will be a variation of u with λ, so,

$$v_g = \frac{d\omega}{dk} = \frac{d\omega}{d\left(2\pi/\lambda\right)}$$

$\Rightarrow$

$$v_g = \frac{d\left(\dfrac{2\pi}{\lambda} u\right)}{-\dfrac{2\pi}{\lambda^2} d\lambda}$$

$\Rightarrow$

$$v_g = -\lambda^2 \frac{d}{d\lambda}\left(\frac{u}{\lambda}\right)$$

$\Rightarrow$

$$v_g = -\lambda^2 \left[\frac{1}{\lambda}\frac{du}{d\lambda} - \frac{u}{\lambda^2}\right]$$

$\Rightarrow$

$$\boxed{v_g = u - \lambda \frac{du}{d\lambda}}$$

In the absence of dispersion, $\dfrac{du}{d\lambda} = 0$, we have, $v_g = u$.

6. DECLINE OF OLD QUANTUM THEORY

It was soon realized that the quantum theory as advanced by Planck and Bohr was not adequate for complete analysis of atomic systems. The main short comings of the old quantum theory are :

(I)   It could not explain the spectral lines of system like Hydrogen molecule and normal helium atom.

(II)   It failed to find any information about the transition probabilities and within the spectral lines and the intensities of spectral lines.

(III) It could not explain the process connected with the electron spin and Pauli's exclusion principle.

(IV)   The hypothetical assumption of Bohr about stationery orbits had no theoretical justifications.

(V)   It failed to explain 1/2 values which were needed to obtain agreement with experiments.

In quest of more complete theory, Schrödinger discovered an equation, which could describe complete behaviour of atomic systems in many cases. Heisenberg, Born, Jordan, Dirac did further refinement of this theory and others and that resulted in the form of new quantum theory.

7. SCHRÖDINGER EQUATION

In wave mechanics particle motion thus described by wave packet. A wave packet comprises of a group of waves with slightly different velocity and consists of slightly different wavelengths. However, the serious difficulty arises with concept of wave packet. Since wave packet dissipates very soon with time, there must be guiding wave with this wave packet which can describe particle motion. Hence, wave packet plus the guiding wave possesses the properties of a particle moving with velocity 'v'. The amplitude of this guiding wave determines the probability of finding a material at a point. If the amplitude of guiding wave is zero at a point in space, the probability of finding the material particle at this point is infinitesimal. Thus wave packet plus the guiding waves possesses the properties of a particle moving with a velocity v. In order to locate the position of the particle Schrödinger derived the equation. There are two Schrödinger equations. One is dependent on time and the other is not depending on time.

(a) Time Independent Schrödinger Equation

We consider a system of stationary waves to be associated with the particle. Let $\psi(x, t)$ be the wave function which is associated with it. This is the wave characteristic for the de-Broglie wave at any location $\vec{r} = \hat{i}x + \hat{j}y + \hat{k}z$ at time t.

We know the wave equation in three dimensions can be expressed as

$$\nabla^2\psi = \frac{1}{u^2}\frac{\partial^2\psi}{\partial t^2} \quad \Rightarrow \quad \frac{\partial^2\psi}{\partial x^2} + \frac{\partial^2\psi}{\partial y^2} + \frac{\partial^2\psi}{\partial z^2} = \frac{1}{v^2}\frac{\partial^2\psi}{\partial t^2} \qquad \ldots(1)$$

where, v be the velocity of the wave.

The solution of eqn. (1) is assumed as

$$\psi\,(\vec{r},\,t) = \psi_0\,(\vec{r})\,e^{-i\omega t} \qquad \ldots(2)$$

where, ψ_0 is the amplitude at the point considered. It is a function of $\vec{r}$, not 't'.

Using eqns. (1) and (2), we have,

$$\frac{\partial^2 \psi}{\partial x^2} + \frac{\partial^2 \psi}{\partial y^2} + \frac{\partial^2 \psi}{\partial z^2} = (-\,i\omega)\,.\,(-\,i\omega)\,.\,\frac{1}{v^2}\,\psi = -\frac{\omega^2}{v^2}\,\psi \qquad \ldots(3)$$

Now, $$\omega = 2\pi v = 2\pi\,.\,\frac{v}{\lambda}$$

$$\Rightarrow \qquad\qquad \frac{\omega}{v} = \frac{2\pi}{\lambda} \qquad \ldots(4)$$

Using (3) and (4),

$$\frac{\partial^2 \psi}{\partial x^2} + \frac{\partial^2 \psi}{\partial y^2} + \frac{\partial^2 \psi}{\partial z^2} = \nabla^2 \psi = -\frac{4\pi^2}{\lambda^2}\,\psi$$

$$\Rightarrow \qquad\qquad \nabla^2 \psi + \frac{4\pi^2}{\lambda^2}\,\psi = 0 \qquad \ldots(5)$$

Using de-Broglie's equation for wave-particle duality as

$$\lambda = \frac{h}{mv}$$ and, putting this into eqn. (5) we have,

$$\nabla^2 \psi + \frac{4\pi^2 m^2 v^2}{h^2}\,\psi = 0 \qquad \ldots(6)$$

If 'E' and 'V' represent the total and potential energies of the particle respectively, then, its kinetic energy is

$$\frac{1}{2}\,mv^2 = E - V$$

$$\Rightarrow \qquad\qquad (mv)^2 = 2m\,(E - V)$$

Using this into eqn. (6), we have,

$$\nabla^2 \psi + \frac{8m\pi^2}{h^2}\,(E - V)\,\psi = 0 \qquad \ldots(7)$$

As $h = \dfrac{h}{2\pi}$, we have from eqn. (7),

$$\nabla^2 \psi + \frac{2m}{h^2}\,(E - V)\,\psi = 0 \qquad \ldots(8)$$

Above eqns. (7) and (8) are known as Time Independent Schrödinger equation.

(b) Time Dependent Schrödinger Equation

We have from eqn. (2), $\psi = \psi_0\,e^{-i\omega t}$

so,
$$\frac{\partial \psi}{\partial t} = (-i\omega)\, \psi_0 \cdot e^{-i\omega t}$$

$$\Rightarrow \qquad \frac{\partial \psi}{\partial t} = -i > (2\pi\nu) \cdot \psi$$

$$\Rightarrow \qquad \frac{\partial \psi}{\partial t} = -2\pi i \cdot \left(\frac{E}{h}\right) \cdot \psi = -\left(\frac{iE}{\hbar}\right)\psi$$

$$\Rightarrow \qquad E\psi = i\hbar\, \frac{\partial \psi}{\partial t} \qquad \qquad \dots(9)$$

Using eqns. (9) into (8), we have,

$$\nabla^2 \psi + \frac{2m}{h^2}\left(i\hbar\, \frac{\partial \psi}{\partial t} - V\psi \right) = 0$$

$$\Rightarrow \qquad -\frac{h^2}{2m}\, \nabla^2 \psi + V\psi = i\hbar\, \frac{\partial \psi}{\partial t}$$

$$\Rightarrow \qquad \left[-\frac{h^2}{2m}\, \nabla^2 + V \right]\psi = -i\hbar\, \frac{\partial \psi}{\partial t} \qquad \dots(10)$$

This eqn. (10) contains time and hence it is known as Time Dependent Schrödinger equation.

The operator $\left[-\dfrac{h^2}{2m}\, \nabla^2 + V \right]$ is called Hamiltonian and is represented by 'H', hence, equation (10) can be written as

$$H\psi = E\psi \qquad \qquad \dots(11)$$

This form of Schrödinger equation represents the motion of a non-relativistic material particle.

8. PHYSICAL SIGNIFICANCE OF WAVE FUNCTION 'ψ'

In the beginning the wave function 'ψ' was considered merely an auxiliary mathematical quantity to facilitate computations relative to the experimental results. But it does not seem reasonable to describe the wave function just as a mathematical function without its physical significance. Schrödinger himself attempted to give interpretation of 'ψ' in terms of charge density. If 'ψ' is the amplitude of the matter wave at any point in space, then according to electromagnetic wave system number of material particle per unit volume, i.e., the particle density must be proportional to ψ^2. Thus square of the absolute value of measures particle density. If q be the electric charge of the particle then $q\,|\psi|^2$ is the charge density as considered by Schrödinger. Usually $\psi^*\psi$ is written as $|\psi|^2$. But this theory could not explain the wave packet associated with the moving particle as wave packet vanishes after a interval of time.

To remove above discrepancy Max Born in 1926 gave its physical significance which is widely accepted till date. According to him

$\psi^*\psi = |\psi|^2$ represents the probability density of the particle in the state 'ψ'. Thus the probability of finding a particle in the volume element dV about any point 'r' at time 't' is $P(r)\, dV = |\psi(r, t)|^2\, dV$.

This postulate suggests that the quantum mechanical laws and the results of their measurements can be interpreted on the basis of probability considerations. Since the particle is present certainly somewhere, so,

$$\int_{-\infty}^{+\infty} |\psi|^2\, dV = 1.$$

A wave function which obeys the above equation is called as normalized. Hence, 'ψ' does not have its own physical significance; however, $|\psi|^2$ has its physical significance.

Properties of 'ψ' :

1. It must be finite everywhere.
2. It must be single valued.
3. It must be continuous and also its first derivative to be continuous.

Postulates of Quantum Mechanics

The mathematical formulation of quantum mechanics is based on the following postulates :

I.   Wave function associated with every physical state of the system contains the entire description. The wave function is a function of all position coordinates and time, and, contains the information about the properties of the system.

II. Every physical observable is associated with an operator.

$$\text{Energy} : \hat{E} = i\hbar\, \frac{\partial}{\partial t} \; ; \; \text{Momentum} : \hat{p} = \frac{\hbar}{t}\, \vec{\nabla}.$$

III. The measurements of an observable can provide the values of λ given by the equation $\hat{p}\, \phi = \lambda\phi$. This equation is known as eigen value equation and the values of λ are called eigen values. Φ is a wave function, only single valued.

9. | ELEMENTARY IDEA OF QUANTUM STATISTICS

There are two types of statistics.

I. Classical (Maxwell-Boltzmann) Statistics

It applies to the particles which are :
(i) Identical,   (ii) Distinguishable,
Examples : Molecules of a gas.

II. Quantum Statistics

There are two types of quantum statistics.

(A) Bose-Einstein (BE) statistics : It applies to the particles which are :

(i)   Identical,

(ii) Indistinguishable,

(iii) Having integral spin (spin 1),

(iv) Particles are known as Bosons.

Examples : Photons in radiation. The Bose-Einstein distribution law is given by

$$n_i = \frac{g_i}{e^{\alpha + \beta\varepsilon_i} - 1}$$

This equation represents the most probable distribution of the particles (n_i) among various energy levels (ε_i) for a system obeying BE statistics. α and β are constants. g_i denotes the degeneracy of ith level.

(B) Fermi-Dirac (FD) statistics : It applies to the particles which are :

(i)   Identical,

(ii) Indistinguishable,

(iii) Having half-integral spin.,

(iv) Particles are known as Fermions.

Examples : Conduction electrons in metals, proton beam, neutron beam, etc. The Fermi-Dirac distribution law is given by

$$n_i = \frac{g_i}{e^{\alpha + b\varepsilon_i} + 1}$$

This equation represents the most probable distribution of the particles (n_i) among various energy levels (ε_i) for a system obeying FD statistics.

3

FREE ELECTRON THEORY

Metal is the most important form of solid. Due to their mechanical strength, conductivity and abundance in nature, they have found various applications in engineering. Therefore these materials are being utilized in the development of society. More investigation have been made from the beginning of twentieth century. An atom is considered to be made up of electrons, protons and neutrons. The electrons, revolving around the nucleus in various orbits play important role in various engineering application. However, all electrons of metal do not participate in various conduction properties. The valence shell electrons of metals are responsible for overall behaviour of metal. These electrons do not affect the motion of core electrons and nucleus of atoms. Various models have been applied to investigate the electrons motion so that controlling parameter for describing the properties of materials under external influence can be defined. Modifications in the previous theory have been made by applying quantum mechanics. Fundamental terms, definition are given in this chapter, which are important tool to understand the functioning of devices. In this chapter valence shell electrons have been treated as free electrons and their motion is subjected under external effect. Fundamental aspects have been given while describing the motion of free electrons through models consisting of various assumptions. Further modifications made by scientist into the theory to meet the experimental results have been discussed. These theoretical models make us to equip with fundamental understanding of nanoelectronics. The confined motion of electrons in certain boundaries has been dealt here, which is helpful to understand the motion of electrons in nanostructure devices.

1. ELEMENTS OF FREE ELECTRON THEORY (CLASSICAL)

In this theory each metal contains a number of free electrons which behave like the molecules of a perfect gas and are free to move about the whole volume of metal. They thus form **free electron gas** in a container. In this model it is assumed that the free electrons, *i.e.*, those giving rise to the conductivity, find themselves in a potential which is constant everywhere inside the metal. Since one does not observe electron emission from metals at room temperature, it seems evident that potential energy of an electron's at rest outside the metal. Based on this model Lorentz and Drude in 1900 put forward their theory for the calculation of conductivity of metals.

Drude Theory (Lorentz – Drude)

The basic **assumptions** of Lorentz – Drude theory are listed below :

(i) The large no. of electrons in a metal is free to move about the whole volume of the metal.

(ii) The motion of free electrons here is similar to the thermal agitation of a perfect gas. The electrical or thermal conductivity of metals is solely due to these electrons.

(iii) The free electron here move randomly in all possible directions with widely different velocities. If v_i is the random velocity of ith electron and n is the total number of electrons, then, $\sum\limits_{i=1}^{n} \overrightarrow{v_i} = 0$.

Also the distribution of velocity is in accordance with Maxwell's distribution.

So, the average kinetic energy (KE) of a free electron is $\dfrac{3}{2}kT$, where, k is Boltzmann's constant and T is absolute temperature of it.

(iv) The free electrons make collisions with positive ions (fixed) in the lattice and also among themselves. These collisions among electrons have no practical contribution to the conductivity of metals.

(v) In absence of electric field random motion in all directions of electrons occur. So current density in metal is zero.

(vi) When the external field is applied the electrons drift slowly with some average velocity known as average drift velocity, in the direction opposite to that of electric field. The drift velocity is superimposed over their random velocity.

(vii) Under external field free electrons being accelerated gain some additional KE and by colliding with fixed positive ion loses its additional KE. Thus there is a loss of KE in collision under electric field (inelastic collision). Just after inelastic collision the electron momentarily will have random motion. Again due to the presence of electric field electron is again accelerated and gains drift velocity to next collision. This process continues till the electric field is acted on.

(viii) The average distance traversed by a free electron between two successive collisions with positive ions is known as Mean Free Path, λ.

(a) **Determination of electrical conductivity, σ, of the metal :** Now, we consider a metal with n number of free electrons per cubic meter. Under an external electric field of E volts/meter, the force experienced by a free electron is $-e\overrightarrow{E}$ Newton.

The gain in momentum after a time t then is $-e\overrightarrow{E} \times t$ Newton-sec.

If m is the mass of the free electron, t_i is time for ith electron, then, the momentum of ith electron at that instant is $(m\overrightarrow{v_i} - e\overrightarrow{E}t_i)$.

The average momentum of electrons at that instant is

$$\overrightarrow{P}_{av} = \frac{\sum\limits_{i=1}^{n} (m\overrightarrow{v_i} - e\overrightarrow{E}t_i)}{n}.$$

So, average drift velocity,

$$\overrightarrow{v_d} = \frac{\overrightarrow{P}_{av}}{m} = \frac{1}{n} \sum_{i=1}^{1} \left(\overrightarrow{v_i} - \frac{e\overrightarrow{E}}{m} t_i\right) \qquad \text{...(1)}$$

Now, the velocities of free electrons are randomly distributed,

$$\Rightarrow \qquad\qquad \sum_{i=1}^{n} \overrightarrow{v_i} = 0 \qquad\qquad \text{...(2)}$$

From eqns. (1) & (2),

$$\overrightarrow{v_d} = \frac{1}{n} \sum_{i=1}^{n} \left(-\frac{e\overrightarrow{E}}{m} t_i\right)$$

$$\Rightarrow \qquad\qquad \overrightarrow{v_d} = -\frac{e\overrightarrow{E}}{m} \frac{\sum\limits_{i=1}^{n} t_i}{n}$$

$$\Rightarrow \qquad\qquad \overrightarrow{v_d} = -\frac{e\overrightarrow{E}}{m} t \qquad\qquad \text{...(3)}$$

Here, $t = \dfrac{1}{n} \sum\limits_{i} t_i$

This is the average time elapsed after collisions. This average time will be equal to the average time from this instant to the next collision. So, if the collisions occur at regular intervals the average time τ between collisions will be, $t = \tau$. Here t is the relaxation time.

So, from eqn. (3),

$$\overrightarrow{v_d} = -\frac{e\overrightarrow{E}}{m} \tau \; ;$$

Hence, the current density,

$$\overrightarrow{J} = -ne\overrightarrow{v_d}$$

[Here, current, $\qquad I = -\dfrac{N_0 e}{t}$;

N_0 is total no. of electron within volume V & t is time rate,

$$I = -n \times V \times \dfrac{e}{t}$$

$\Rightarrow \qquad\qquad\qquad I = -n \times \dfrac{L \times A}{t} \times e$

$\Rightarrow \qquad\qquad\qquad J = -\dfrac{I}{A} = -n \times \dfrac{L}{t} \times e = -nv_d e]$

So, $\qquad\qquad\qquad \vec{J} = -ne \times -\dfrac{e\vec{E}}{m} \times \tau = \dfrac{ne^2 \tau}{2m} \vec{E}$

$\Rightarrow \qquad\qquad\qquad \left| \vec{J} \right| = \dfrac{ne^2 \tau}{m} \vec{E}$

But, $\vec{J} = \sigma \vec{E}$; σ is electrical conductivity of the metal,

Hence, comparing above two expressions of $\vec{J}$,

$$\sigma = \dfrac{ne^2 \tau}{m} \qquad\qquad ...(I)$$

From equation (I) we found that the differences in σ for different metals are due to the difference in n. A metal is pictured as composed of positive metal ions whose valence electrons are free to move around in three dimension ionic array. As there are 10^{22} to 10^{23} atom/cc, a large number of free electrons is available moving in all directions inside the metal like the atoms or molecules of a perfect gas.

(b) Determination of thermal conductivity, K, of the metal : Based on the assumptions of this metal the thermal conductivity can also be calculated. Let us consider a rectangular bar, as shown in Fig. 1 given below

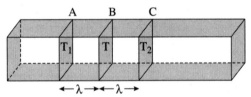

Fig. 1 A rectangular metal bar showing three different slices at different temperatures.

in which thermal energy is flowing in positive X-direction. Let A, B, C are three slices of this bar having uniform cross-section. Due to the existence of thermal gradient the temperatures of the slices A, B, C are $T_1\, T$ and T_2 respectively in such a way that $T_1 > T > T_2$. Let λ be the mean free path of the electrons separating slices A, B, C, then, according to Joule's law, the heat flux is

$$\Delta Q \propto \dfrac{\Delta T}{\Delta x}$$

$$\Rightarrow \qquad \qquad \Delta Q = K \frac{\Delta T}{\Delta x} = K\left[\frac{T_1 - T_2}{2\lambda}\right] \qquad \qquad ...(1)$$

Equation (1) gives the amount of energy per unit area per unit time exchanged in the slice B.

Let n be the number of electrons per unit volume, then $n/6$ will be the number of electrons per unit volume moving in positive X-direction. If v is the average velocity of the electrons, then, $nv/6$ will represent the number of electrons flowing in positive X-direction per unit area per unit time. Thus the amount of energy transfer from the slice A to B per unit area per unit time is $Q_1 = (nv/6) \times \frac{3}{2} kT_1$.

Similarly the amount of energy transfer from the slice B to C per unit area per unit time is $Q_2 = (nv/6) \times \frac{3}{2} kT_2$. Here, k represents the Boltzmann's constant.

So net heat flux exchange in B is

$$\Delta Q = Q_1 - Q_2 = (nv/6) \times \frac{3}{2} k\,(T_1 - T_2) = (nv/4)\,k\,(T_1 - T_2) \qquad ...(2)$$

Hence by equating eqns. (1) and (2),

$$K\left[\frac{T_1 - T_2}{2\lambda}\right] = (nv/4)\,k\,(T_1 - T_2)$$

$$\Rightarrow \qquad \qquad K = \frac{1}{2} nvk\lambda \qquad \qquad ...(II)$$

From equation (II) we found that the differences in K for different metals are due to the difference in n.

(c) **Wiedemann-Franz law : Statement :** The ratio of thermal conductivity, K, to electrical conductivity, σ, is proportional to the absolute temperature and this constant of proportionality is known as 'Lorentz Number'.

Proof : We have for a free electron gas from eqn. (II),

$$K = \frac{1}{2} n \kappa \lambda v$$

For a particle with v and $\quad \lambda, \tau = \dfrac{\lambda}{v}$ $\qquad \qquad ...(1)$

Also from eqn. (I),

$$\sigma = \frac{ne^2}{m}\,\tau = \frac{ne^2}{m}\frac{\lambda}{v} \qquad \qquad ...(2)$$

Now, KE in three dimension is $\dfrac{1}{2} mv^2$ and equals to $3 \times \dfrac{1}{2} \kappa T$.

$$\therefore \qquad \qquad \frac{1}{2} mv^2 = 3 \times \frac{1}{2} \kappa T$$

$$\Rightarrow \qquad\qquad mv = \frac{3\,\kappa T}{v}$$

$$\Rightarrow \qquad\qquad mv = \frac{3\,\kappa T}{v} \qquad\qquad\qquad ...(3)$$

From eqns. (2) and (3),

$$\sigma = \frac{ne^2}{m}\frac{\lambda}{v} = \frac{ne^2\lambda}{1}\times\frac{1}{3\,\kappa T/v} = \frac{ne^2\lambda v}{3\,\kappa T} \qquad\qquad ...(4)$$

Using eqns. (II) and (4), $\qquad \dfrac{K}{\sigma} = \dfrac{\dfrac{1}{2}\,n\,\kappa\,\lambda\,v}{ne^2\lambda v/3\,\kappa T} = \dfrac{3\,\kappa^2}{2e^2}\,T$

$$\Rightarrow \qquad\qquad \boxed{\frac{K}{\sigma} \propto T = LT}$$

Here, $L = \dfrac{K}{\sigma T} \Rightarrow$ This is known as Lorentz number and is given by

$$L = \frac{3K^2}{2e^2}$$

2. | LIMITATIONS OF FREE ELECTRON THEORY (FET)

(1) **Specific Heat :** The electronic contribution of specific heat has serious disagreement with the experimental results. The FET is unable to explain the reason of dominance of low temperature specific heat evaluation by its electronic contribution.

(2) **Paramagnetism of Metals :** Experimental fact that the paramagnetism of Metals is nearly independent of temperature, this is not at all explained by Free Electron theory.

(3) **Widemann-Franz Law :** This is a consequence of Free Electron Theory. It gives the value of $\dfrac{k}{\sigma T}$, here, k is the thermal conductivity of the metal concerned, which are found to agree with theoretical value at ordinary temperature, but deviations occur at low temperature.

(4) **Hall Coefficient :** Free electron theory predicts the value of hall Coefficient being independent of temperature, relaxation time or the strength of magnetic field applied. This is not at all found experimentally.

(5) **The magneto resistance :** FET predicts that the resistance of a wire perpendicular to a uniform magnetic field should not dependent on the strength of the field. Practically it does.

(6) **Temperature dependence of Electrical Conductivity :** FET does not predict. However, it practically exists.

(7) **Directional dependence of Electrical Conductivity :** For some metals σ depends on the orientation of the specimen (if suitably prepared) with respect to the electric field. In such specimens $\vec{J}$ need not even be parallel to $\vec{E}$.

3. QUANTUM THEORY OF CONDUCTION

The modification of FET was done by using quantum statistics (by Sommerfield, 1928) on the basis of the following assumptions.

(i) The large numbers of conduction electrons in metals are not completely free. They are bound to the metal as a whole. We now arrive a physical model in which the interior of the metal is represented by a potential energy (PE) box of depth E_S as shown in Fig. 2 given next. E_S is the energy difference between an

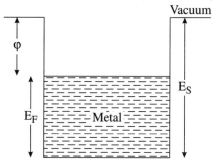

Fig. 2   A PE box-depicting a metal.

electron at rest inside the metal and in vacuum. At $T = 0$ K, all energy levels up to E_F (known as Fermi energy) are completely filled and all the higher ones are empty. Here, φ, known as the work function of the metal and is the work required to be done to extract an electron from metal. This is given by, $\phi = E_S - E_F$.

(ii) The forces between conduction electrons and ion cores are neglected, so that the electron within the metal is treated as free. The total energy of an electron is wholly the kinetic energy (KE) as PE is negligible due to their smaller mass.

(iii) Due to their light mass and dense packing the electrons in a metal may be considered as a gas under very high compression and hence to a degenerate gas. Moreover, as the electron gas is charged, the free electron gas in metal may be considered as dense Plasma.

(iv) The electrons are assumed to obey Pauli's exclusion principle, so they follow the Fermi-Dirac (FD) statistics, rather than the classical Maxwell-Boltzmann (M-B) statistics.

4. SOLUTION OF ONE DIMENSIONAL SCHRÖDINGER EQUATION IN A CONSTANT POTENTIAL

We now consider a single particle confined in a region of constant potential V_0. It is one dimensional motion is governed by the region $0 < x < L$ as shown in Fig. 3.

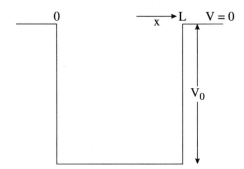

Fig. 3 A potential well of depth V_0.

The probability of finding the particle outside this region is zero. The wave function $\Psi(x, t)$ describes the behaviour of the particle.

So, $\Psi(x, t) = 0$ for x excluding $0 < x < L$.

The potential function V_0 is independent of time t, we can then write one dimensional Schrödinger equation of the particle as

$$\frac{d^2 \Psi(x)}{dx^2} + \frac{2m}{\hbar^2} (E - V_0) \Psi(x) = 0 \qquad \ldots(1)$$

[as V_0 is constant we are only considering Ψ to be a function of x and independent of t]

The solution of eqn. (1) is

$$\Psi(x) = A \sin \left(\sqrt{\frac{2m}{\hbar^2} (E - V_0)} \right) x + B \cos \left(\sqrt{\frac{2m}{\hbar^2} (E - V_0)} \right) x \quad \ldots(2)$$

Here, A and B are integral constants.

Now, for this problem the boundary conditions are

$$\left. \begin{array}{c} \Psi(x) = 0 \text{ at } x = 0 \\ \Psi(x) = 0 \text{ at } x = L \end{array} \right\} \qquad \ldots(3)$$

and

Using the first part of eqn. (3) in eqn. (2) we get,

$$\Rightarrow \qquad 0 = 0 + B \times 1 \quad \Rightarrow \quad B = 0$$

So, eqn. (2) now modifies into

$$\Psi(x) = A \sin kx \qquad \ldots(4)$$

where,

$$k = \left(\sqrt{\frac{2m}{\hbar^2} (E - V_0)} \right) \qquad \ldots(5)$$

Also using the second part of eqn. (3) in eqn. (4) we get,

$$0 = A \sin (kL)$$

$$\Rightarrow \qquad A = 0$$

This is **not possible** as both A and B cannot become zero.

Hence, $\qquad \sin (kL) = 0 = \sin (n\pi) \qquad$ (where, $n = 0, \pm 1, \pm 2, \ldots$)

$$\Rightarrow \qquad k^2 = \left(\frac{n\pi}{L} \right)^2$$

$$\Rightarrow \qquad \frac{2m}{\hbar^2}(E - V_0) = \left(\frac{n\pi}{L}\right)^2$$

$$\Rightarrow \qquad E - V_0 = \frac{n^2\pi^2\hbar^2}{2mL^2} \qquad \qquad ...(6)$$

In eqn. (6), $n = 0$

$\Rightarrow$ No value for $(E - V_0)$, *i.e.*, $\Psi(x)$ does not exist $\Rightarrow$ This is not possible.

Also, for $n = -1, -2, ...$ $\Rightarrow$ same value of $(E - V_0)$ as for $n = 1, 2, 3, ...$

Hence we finally get,

$$E - V_0 = \frac{n^2\pi^2\hbar^2}{2mL^2} \quad \text{where } n = 1, 2, 3, ... \quad ...(7)$$

Thus the particle can only have **only certain discrete energy values** given by

$$E = V_0 + \frac{n^2\pi^2\hbar^2}{2mL^2} \quad \text{where } n = 1, 2, 3, ... \qquad ...(8)$$

These are known as **energy eigen values.**

Now, $k = \dfrac{n\pi}{L}$, so from eqn. (4),

$$\Psi_n(x) = A \sin\frac{n\pi}{L}x \qquad \qquad ...(9)$$

The value of A can be obtained from the normalization condition of $\Psi_n(x)$, *i.e.*,

$$\int_{-\infty}^{+\infty} \Psi^* \Psi = 1$$

$$\Rightarrow \qquad \int_{-\infty}^{+\infty} |\Psi_n(x)|^2 \, dx = 1$$

$$\Rightarrow \qquad \int_0^L |\Psi_n(x)|^2 \, dx = 1 \qquad (\text{as } \Psi_n(x) = 0 \text{ outside } 0 < x < L)$$

$$\Rightarrow \qquad \int_0^L \left| A \sin\frac{n\pi x}{L} \right|^2 dx = 1$$

$$\Rightarrow \qquad |A|^2 \int_0^L \sin^2\left(\frac{n\pi x}{L}\right) dx = 1$$

$$\Rightarrow \qquad |A|^2 \times \frac{L}{2} = 1 \implies |A| = \sqrt{\frac{2}{L}}$$

$$\Rightarrow \qquad A = \sqrt{\frac{2}{L}}$$

Hence,

$$\Psi_n(x) = \sqrt{\frac{2}{L}} \sin \frac{n\pi}{L} x \qquad \qquad ...(10)$$

Ground state wave function is

$$\Psi_1(x) = \sqrt{\frac{2}{L}} \sin \frac{\pi}{L} x.$$

First excited state is $\Psi_2(x) = \sqrt{\frac{2}{L}} \sin \frac{2\pi}{L} x$, and so on. They are shown in adjoining Fig. (4). For three dimensional case the electrons are

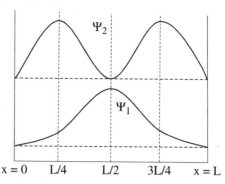

Fig. 4 Energy states in potential well.

confined into a cube of volume L^3 and so eigen functions are

$$\Psi(x, y, z) = \sqrt{\frac{8}{L^3}} \sin\left(\frac{n_x \pi x}{L}\right) \sin\left(\frac{n_y \pi y}{L}\right) \sin\left(\frac{n_z \pi z}{L}\right) \text{ and eigen values are}$$

$$E_n (n_x, n_y, n_z) = \frac{\pi^2 \hbar^2}{2mL^2} (n_x^2 + n_y^2 + n_z^2)$$

where,

$$k_x = \frac{n_x \pi}{L}, k_y = \frac{n_y \pi}{L}, k_z = \frac{n_z \pi}{L}.$$

Here, n_x, n_y and n_z are integers, having values 1, 2, 3, ... and $k_x = 0, \pm \frac{2\pi}{L}, \pm \frac{4\pi}{L}, ...$ etc. and also same for k_y and k_z.

4A. PARTICLE IN A BOX

Free particle moving in a one dimensional box : When a free particle is moving inside the infinitely high potentially barrier particle is confined to move inside this barrier as shown in figure 5. If we consider particle's motion in one dimension then it may be treated as a free particle moving in a one dimensional box.

The probability of finding the particle outside this region is zero as

$$V = \infty \quad \text{for} \quad 0 > x > L$$

and

$$V = 0 \quad \text{for} \quad 0 < x < L$$

The wave function $\Psi(x, t)$ describes the behaviour of the particle.

So, $\Psi(x, t) = 0$ for x having $0 > x > L$.

We can then write one dimensional Schrödinger equation of the particle within this box as

$$\frac{d^2\Psi(x)}{dx^2} + \frac{2mE}{\hbar^2}\psi(x) = 0 \qquad ...(1)$$

The solution of eqn. (1) is

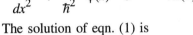

Fig. 5 Particle in a one dimensional box.

$$\Psi(x) = A\sin\left(\sqrt{\frac{2mE}{\hbar^2}}\right)x + B\cos\left(\sqrt{\frac{2mE}{\hbar^2}}\right)x \qquad ...(2)$$

Here, A and B are integral constants.

Now, for this problem the **boundary conditions** are

$$\Psi(x) = 0 \text{ at } x = 0 \text{ and } \Psi(x) = 0 \text{ at } x = L \qquad ...(3)$$

Using the first part of eqn. (3) in eqn. (2) we get,

$$\Rightarrow \qquad\qquad 0 = 0 + B \times 1 \quad \Rightarrow \quad B = 0$$

So, eqn. (2) now modifies into

$$\Psi(x) = A\sin kx \qquad ...(4)$$

where,

$$k = \left(\sqrt{\frac{2mE}{\hbar^2}}\right) \qquad ...(5)$$

Also using the second part of eqn. (3) in eqn. (4) we get,

$$0 = A\sin(kL)$$

$$\Rightarrow \qquad\qquad A = 0$$

This is **not possible** as both A and B cannot become zero to have finite Ψ.

Hence, $\qquad\qquad \sin(kL) = 0 = \sin(n\pi)$ (where, $n = 0 \pm 1, \pm 2, ...$)

$$\Rightarrow \qquad\qquad k^2 = \left(\frac{n\pi}{L}\right)^2$$

$$\Rightarrow \qquad\qquad \frac{2mE}{\hbar^2} = \left(\frac{n\pi}{L}\right)^2$$

$$\Rightarrow \qquad\qquad E = \frac{n^2\pi^2\hbar^2}{2mL^2} \qquad ...(6)$$

In eqn. (6), $n = 0$.

$\Rightarrow$ No value for E, *i.e.*, $\Psi(x)$ does not exist $\Rightarrow$ This is not Possible.

Also, for $n = -1, -2, ... \Rightarrow$ same value of E as for $n = 1, 2, 3, ...$

Hence we finally get,

$$E = \frac{n^2\pi^2\hbar^2}{2mL^2} \qquad \text{where } n = 1, 2, 3, ... \quad ...(7)$$

Thus the particle can only have **only certain discrete energy values** given by

$$E = \frac{n^2\pi^2\hbar^2}{2mL^2} \qquad \text{where } n = 1, 2, 3, \dots \quad \dots(8)$$

5. DENSITY OF STATES

We have seen that the energy levels of a particle in a region of constant potential are quantized. The number of energy levels per unit energy range per unit volume at a given energy is called the **Density of States**. This is represented by DOS, or $g(E)\, dE$, and is given by $\frac{dn}{dE_n}$.

Each energy band in a crystal accommodates a large numbers of electron energy levels. Each energy level consists of two states and each state accommodates only one electron. One energy level can be occupied by two electrons only having opposite directions of spin. DOS represents the number of states that can be occupied by charge carriers. However, all the available energy states are not filled in an energy band. A particular energy level E is occupied or not is determined by the probability $f(E)$ (this is known as the Fermi-Dirac distribution function as the electrons are Fermions) that a carrier can have the energy E.

If $G(E)\, dE$ is the number of available quantum states in the energy range between E and $E + dE$, then the carrier concentration in energy range dE is obtained by

$$dN = N(E)\, dE = f(E)\, G(E)\, dE \qquad \dots(1)$$

where, $N(E)$ is the carrier distribution function.

In order to calculate the DOS we consider a solid sphere in n point space as shown in Fig. (6) below. There any point (n_x, n_y, n_z) with integer values of co-ordinates represent an energy state. Thus, all the points on the surface of the sphere of radius n (where, $n^2 = n_x^2 + n_y^2 + n_z^2$) will have the same energy. Since n_x, n_y, n_z can have only positive integral value, so, the number of states of energy E will be the octant of the sphere, *i.e.*,

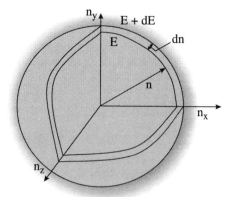

Fig. 6 A point sphere of radius n.

$$G(E) = \frac{1}{8}\left(\frac{4}{3}\,\pi n^3\right).$$

We know from earlier, $E = \dfrac{n^2 \pi^2 \hbar^2}{2mL^2} \Rightarrow n = \dfrac{L}{\pi \hbar} [2mE]^{1/2}.$

Hence, $\qquad G(E) = \dfrac{1}{8} \cdot \dfrac{4\pi}{3} \left(\dfrac{2mL^2}{\pi^2 \hbar^2} \right)^{3/2} \cdot E^{3/2}$ ...(2)

Now, on differentiation,

$$G(E)\, dE = \dfrac{1}{8} \cdot \dfrac{4\pi}{3} \left(\dfrac{2mL^2}{\pi^2 \hbar^2} \right)^{3/2} \dfrac{3}{2} \cdot E^{1/2} \cdot dE$$

Simply, $\quad G(E)\, dE = \dfrac{1}{8} \cdot \dfrac{4\pi}{3} \left(\dfrac{2mL^2}{\pi^2 \hbar^2} \right)^{3/2} \dfrac{3}{2} \cdot E^{1/2} \cdot dE$

Using Pauli's exclusion principle, the number of energy states lying between E and $E + dE$ in volume V $(= L^3)$ is

$$G(E)\, dE = 2 \times \dfrac{1}{8} \cdot \dfrac{4\pi}{3} \left(\dfrac{2mL^2}{\pi^2 \hbar^2} \right)^{3/2} \cdot E^{1/2} \cdot dE$$

$$= \dfrac{4\pi V}{h^3} (2m)^{3/2}\, E^{1/2} \cdot dE$$

So, DOS, the number of energy states lying between E and $E + dE$ per unit volume V is given by

$$g(E)\, dE = \dfrac{4\pi}{h^3} (2m)^{3/2}\, E^{1/2} \cdot dE \qquad \text{...(3)}$$

Adjoining Fig. (6) shows the variation of DOS as a function of energy in accordance to eqn. (2).

Density of States (three dimension) : For three dimension we consider the linear momentum, $\vec{p}$, which corresponds to the operator, $\vec{p} = -i\hbar \vec{\nabla}$, in quantum mechanics.

Thus for the energy state,

$$p \Psi_k(\vec{r}) = -i\hbar \vec{\nabla}\, \psi_k(\vec{r}) = \hbar \kappa\, \Psi_k(\vec{r}) \qquad [\because p = \dfrac{h}{\lambda} = \dfrac{h}{2\pi} \dfrac{2\pi}{\lambda} = \hbar \kappa]$$

So, the plane wave Ψ_k is an eigen function of the linear operator p with the eigen value $\hbar k$. Hence the particle velocity in the energy state of orbital k is given by $v = \dfrac{\hbar k}{m}$. In the ground state of a system of N free electrons the occupied orbitals may be represented as points inside a sphere in k-space. The energy at the surface of the sphere is E_F and the wave vectors at the Fermi surface have a magnitude k_F as in Fig. 7.

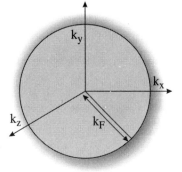

Fig. 7 Fermi surface in k-spaces.

So,
$$E_F = \frac{\hbar^2 k^2}{2m} \qquad \qquad \dots(1)$$

Now the volume of each orbital is $\left(\dfrac{2\pi}{L}\right)^3$. Also the volume of the Fermi surface or sphere of radius k_F is $\dfrac{4\pi}{3}(k_F)^3$. Hence, the total number of orbitals is

$$N = 2 \times \frac{\dfrac{4\pi}{3}(k_F)^3}{(2\pi/L)^3} \qquad \qquad \dots(2)$$

Here '2' comes from spin degeneracy.

So,
$$N = 2 \times \frac{4\pi}{3} \times \frac{k_F^3 \times L^3}{2\pi \times 4\pi^2} = \frac{V}{3\pi^2} k_F^3 \ [V = L^3]$$

$$\Rightarrow \qquad k_F = \left[\frac{3\pi^2 N}{V}\right]^{1/3} \qquad \qquad \dots(3)$$

where, 'N' also denotes the total number of electrons. So, k_F mainly depends upon N/V or particle concentration and not on the mass.

So,
$$E_F = \frac{\hbar^2}{2m}\left[\frac{3\pi^2 N}{V}\right]^{2/3}$$

$$\Rightarrow \qquad N = \frac{V}{3\pi^2}\left[\frac{2mE_F}{\hbar^2}\right]^{3/2} \qquad \qquad \dots(4)$$

The electron velocity v_F at the Fermi surface is

$$v_F = \frac{\hbar k_F}{m} = \frac{\hbar}{m}\left[\frac{3\pi^2 N}{V}\right]^{1/3} \qquad \qquad \dots(5)$$

The density of states now can be expressed as

$$\int_0^{E_F} D(E)\, dE = N$$

$$\Rightarrow \qquad \int_0^{E_F} D(E)\, dE = \frac{V}{3\pi^2}\left[\frac{2mE_F}{\hbar^2}\right]^{3/2} \qquad \text{[from (4)]}$$

$$\Rightarrow \qquad \int D(E)\, dE = \frac{V}{3\pi^2}\left[\frac{2mE}{\hbar^2}\right]^{3/2}$$

[expressing in indefinite integral form]

On differentiation,

$$D(E) = \frac{V}{2\pi^2}\left(\frac{2m}{\hbar^2}\right)^{3/2} E^{1/2}$$

$$= \frac{4\pi^2}{h^3}(2m)^{3/2} \cdot E^{1/2} \qquad \qquad \dots(6)$$

6. FERMI-DIRAC (FD) DISTRIBUTION FUNCTION

We know, $\qquad\qquad$ $N(E)\,dE = f(E)\,G(E)\,dE$

Here, $f(E)$ determines the carrier occupancy of the energy states (*i.e.*, if 100 states has to be shared by 10 carriers then $f(E) = 10/100 = 0.1$. The expression that governs the distribution of electrons among the energy levels as a function of temperature is known as Fermi-Dirac (FD) Distribution function $f(E)$ and, is given by

$$f(E) = \frac{1}{e^{(E-E_F)/\kappa T} + 1} \qquad \dots(1)$$

- $f(E)$ generally indicates the probability of an energy level E to be occupied by an electron.
- $f(E)$ is independent of DOS.
- $f(E) = 1$ if an energy state is filled by electrons.
- $f(E) = 0$ if an energy state is empty.

E_F **is called the Fermi Energy or Fermi Level.**

6A. VARIATION OF $f(E)$ WITH TEMPERATURE

Case-I : Conductor :

(a) At $T = 0$: We have, $f(E) = \dfrac{1}{e^{(E-E_F)/\kappa T} + 1}$.

(i) When $E < E_F$: $E - E_F$ = negative, and hence,

$$f(E) = \frac{1}{e^{-\infty} + 1} = \frac{1}{0+1} = 1.$$

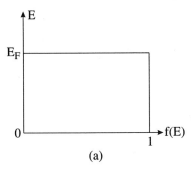
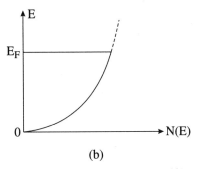

$\qquad\qquad$ (a) $\qquad\qquad\qquad\qquad\qquad\qquad$ (b)

Fig. 8   (a) Variation of energy with FD distribution function for metal at $T = 0°$K
                (b) Variation of no. of energy states with energy.

So, $f(E) = 1 \Rightarrow$ all the levels are occupied by electrons.

The variations of $f(E)$ and $N(E)$ with E are shown in Fig. 8(a) & 8(b) above.

(ii) When $E > E_F$: $E - E_F$ = positive, and hence,

$$f(E) = \frac{1}{e^{+\infty} + 1} = \frac{1}{\infty} = 0$$

So, $f(E) = 0 \Rightarrow$ all the levels above E_F are vacant as shown in above Fig.

(iii) When $E = E_F$: $E - E_F = 0$, and hence,

$$f(E) = \frac{1}{e^{0/0} + 1} \Rightarrow \text{Indeterminable} \Rightarrow \text{Varying between 0 and 1 values.}$$

(b) At $T > 0\,K$: At $T > 0\,K$ electrons are excited to vacant levels above E_F (in general $E_F \gg kT$). Thus a few levels within the range kT below E_F become partially depleted and a few levels within the energy range kT above E_F are partially filled, but most of the deeper levels remain as they were at

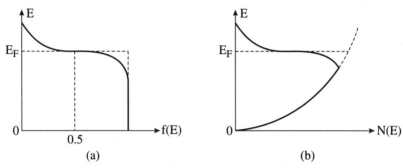

Fig. 9 (a) Fermi distribution function at various energy levels for semiconductor
(b) No. of states variation with energy for semiconductor.

$0\,K$. As a result of thermal excitation the probability of finding electrons at levels $E > E_F$ becomes greater than zero and the probability at levels $E < E_F$ becomes less than unity. The complete distribution function is shown in Fig. 9(a) & 9(b) above.

(i) At $T > 0\,K$, $E = E_F$:

$$f(E) = \frac{1}{e^0 + 1} = \frac{1}{1 + 1} = 0.5$$

- Hence, the probability of occupancy of E_F at $T > 0\,K$ is always 0.5.
- E_F is the average energy possessed by the electrons which participate in conduction process in conductors at $T > 0\,K$. The point where $f(E) = 0.5$ is known as the **Crossover point**. At $E \gg E_F$,

$$f(E) = \frac{1}{e^{E - E_F/\kappa T} + 1} \approx \frac{1}{e^{E - E_F/\kappa T}} = e^{-(E - E_F)/\kappa T}$$

$\Rightarrow$ This is the Maxwell Boltzmann Distribution only.

Case-II : Insulator :

The concept of E_F can be extended to insulator also. As the energy levels in the valence band (VB) are filled $f(E)$ is equal to unity throughout the valence band. As there are no electrons in the conduction band (CB), $f(E)$ is zero throughout the CB. Fig. 10(a) below shows the diagram for an insulator for the probability function.

Case-III : Semiconductor :

A semiconductor is characterized by a CB and VB separated by a smaller energy gap. At normal temperature a significant no. of electrons are thermally excited from VB to CB and an equal numbers of holes are produced in the VB. Fig. 10(b) aside depicts the FD distribution function of the semiconductor. Since the probability of electron occupancy of the CB

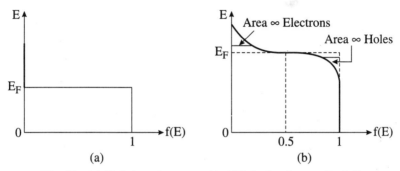

Fig. 10 (a) Variation of energy with $f(E)$ for insulator at $T = 0\ K$
(b) Variation of energy with $f(E)$ for semiconductor at $T > 0\ K$.

enhances at $T > 0\ K$, $f(E)$ is blurred and tapers off towards higher E values in CB. Same situation occurs for holes near the top edge of VB. The extent of blurring of probability curve in both the bands is equal, which indicates that the concentrations of electrons and holes are equal. Also, $f(E) \to 0$, when E increases. Electrons in the CB are clustered very close to the bottom edge of the band. Same is for holes in VB at the top edge of the band. So the electrons are likely to be located at the bottom edge of CB and holes are at top edge of VB. The E_F lies at the middle of the energy gap as it represents the average energy of carriers participating in conduction.

(i) Variation of $f(E)$ with temperature : This variation is clearly depicted in the Fig. 11 given aside. The more the T the more the electrons and holes will be free for conduction.

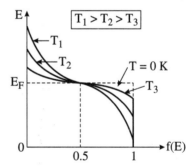

Fig. 11 Variation of energy with $f(E)$ for semiconductor at different temperatures.

7. | CHARACTERISTIC OF FERMI LEVEL (E_F)

(i) E_F is the uppermost occupied energy level in a conductor at $0\ K$ (or maximum energy of a metallic electron at $0\ K$).

(ii) E_F is the average energy possessed by electrons participating in the electrical conduction.

(iii) E_F is a reference level. It is totally occupied in metals.

(iv) When the temperature of a metal is increased only those levels in the vicinity of E_F is disturbed. Some of the levels lower to E_F are vacated and some of the levels above to it are occupied by electrons.

(v) E_F lies in the forbidden gap in case of semiconductors and insulators. This is not occupied here. For semiconductor, fermi level occupancy is 1/2.

7A. FERMI ENERGY AND MEAN ENERGY AT ABSOLUTE ZERO TEMPERATURE

At $T = 0\,K$, when $E < E_F$: $E - E_F$ = negative, and hence,

$$f(E) = \frac{1}{e^{-\infty} + 1} = \frac{1}{0 + 1} = 1$$

Then, the total number of electrons is

$$\int N(E)\, dE = \frac{4\pi V}{h^3} (2m)^{3/2} \int_0^{E_F} E^{1/2}\, dE \qquad \dots(1)$$

(by combining the expressions of $f(E)$ and $G(E)\, dE$).

So, from equation (1),

$$N = \frac{4\pi V}{h^3} (2m)^{3/2} \cdot \frac{2}{3} \cdot (E_F)^{3/2}$$

The number of electrons per unit volume is then given by

$$n = \frac{N}{V} = \frac{4\pi}{h^3} (2m)^{3/2} \cdot \frac{2}{3} \cdot (E_F)^{3/2}$$

$$\Rightarrow \qquad E_F = \frac{h^2}{2m} \left(\frac{3n}{8\pi} \right)^{2/3} \qquad \dots(A)$$

The mean energy at absolute zero temperature is given by

$$\overline{E} = \frac{1}{N} \int_0^{E_F} E N(E)\, dE$$

So, $\qquad \overline{E} = \frac{4\pi V}{Nh^3} (2m)^{3/2} \cdot \int_0^{E_F} E^{3/2}\, dE$

$$\overline{E} = \frac{2}{5} \times \frac{4\pi V}{Nh^3} (2m)^{3/2} \cdot (E_F)^{3/2}$$

$$\Rightarrow \qquad \overline{E} = \frac{\dfrac{2}{5} \times \dfrac{4\pi V}{h^3} (2m)^{3/2} \cdot (E_F)^{5/2}}{\dfrac{4\pi V}{h^3} (2m)^{3/2} \cdot \dfrac{2}{3} \cdot (E_F)^{3/2}}$$

$$\Rightarrow \qquad \overline{E} = \frac{3}{5} E_F$$

8. EFFECT OF TEMPERATURE ON FERMI ENERGY DISTRIBUTION FUNCTION

With the increase in temperature the electrons below Fermi level gains energy and get excited. They then occupy the higher levels which were vacant earlier at absolute zero. The number of free electrons lying in the energy interval 'dE' at any temperature more than the absolute zero is given by

$$N = \int_0^\infty N(E)\, dE = \int_0^\infty g(E) f(E)\, dE$$

$$\Rightarrow \qquad N = \frac{4\pi V}{h^3} (2m)^{3/2} \int_0^\infty \frac{E^{1/2}\, dE}{\exp\left[\dfrac{E - E_F}{kT}\right] + 1} \qquad \ldots(1)$$

Evaluating this integral by integration by parts, we have

$$I = \int_0^\infty \frac{E^{1/2}\, dE}{\exp\left[\dfrac{E - E_F}{kT}\right] + 1}$$

$$\Rightarrow I = \left|\frac{2E^{3/2}}{3} \times \frac{dE}{\exp\left[\dfrac{E - E_F}{kT}\right] + 1}\right|_0^\infty + \frac{2}{3} \int_0^\infty \frac{E^{3/2} \exp\left[\dfrac{E - E_F}{kT}\right] dE}{\left[\exp\left[\dfrac{E - E_F}{kT}\right] + 1\right]^2 . kT}$$

$$\ldots(2)$$

The probability of finding an electron for both (zero and infinite energies) are zero, so, the first term of RHS of equation (2) is zero for both the limits. The second term can be evaluated by using Taylor's series. According to it any function $G(E)$ in the neighborhood of $E = E_F$ can be expanded I powers of $(E - E_F)$ as

$$G(E) = G(E_F) + (E - E_F)\, G\,(E_F) + \frac{(E - E_F)^2}{2!}\, G(E_F) + \ldots$$

This gives us,

$$E^{3/2} = E_F^{3/2} + (E - E_F) . \frac{3}{2} E_F^{1/2} + \frac{(E - E_F)^2}{2!} \frac{3}{4} E_F^{-1/2} + \ldots \quad \ldots(3)$$

Substituting equation (3) into (2), we have

$$I = \frac{2}{3kT} \int_0^\infty \frac{\exp\left[\dfrac{E - E_F}{kT}\right] dE}{\left[\exp\left[\dfrac{E - E_F}{kT}\right] + 1\right]}$$

$$\left[E_F^{3/2} + \frac{3}{2}(E - E_F) . E_F^{1/2} + \frac{3}{8}(E - E_F)^2 . E_F^{-1/2} + \ldots\right] dE \quad \ldots(4)$$

Let we put, $x = \dfrac{E - E_F}{kT}$, so that, $dE = kT . dx$. Also at low temperature $kT \ll E_F$, the derivative $F'(E)$ is large only at energies near $E = E_F$ as shown in Fig. 12.

Here, $$F(E) = \frac{1}{\exp\left[\dfrac{E - E_F}{kT}\right] + 1} \, ;$$

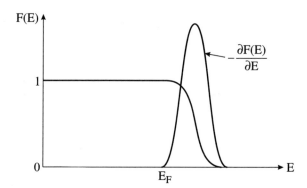

Fig. 12 Energy derivative of fermi distribution function and fermi distribution at $T > 0\,K$.

And
$$F'(E) = \frac{dF(E)}{dE}$$

$$\Rightarrow \qquad F'(E) = -\frac{\left[\exp\left[\dfrac{E-E_F}{kT}\right]+1\right]^{-2}}{kT} \cdot \exp\left[\dfrac{E-E_F}{kT}\right]$$

$$\Rightarrow \qquad F'(E) = -\left(\frac{1}{kT}\right)\frac{\exp\left[\dfrac{E-E_F}{kT}\right]}{\left[\exp\left[\dfrac{E-E_F}{kT}\right]+1\right]^{2}}$$

For other values of E, especially for negative E, $F'(E)$ is negligible. Therefore the lower limit of the integral may be taken as '$-\infty$' instead of $(-E_F/kT)$. The above integral now becomes

$$I = \frac{2}{3kT}\int_{-\infty}^{\infty}\frac{e^x}{(e^x+1)^2}\left[E_F^{3/2}+\frac{3}{2}\,kTxE_F^{1/2}+\frac{3}{8}\,(kTx)^2 \cdot E_F^{-1/2}+\ldots\right]kT\,dx$$

$$\Rightarrow \quad I = \frac{2}{3}\left[E_F^{3/2}\int_0^{\infty}\frac{e^x\,dx}{(e^x+1)^2}+\frac{3}{2}\,kTE_F^{1/2}\int_{-\infty}^{\infty}\frac{xe^x\,dx}{(e^x+1)^2}+\right.$$
$$\left.\frac{3}{8}\,(kT)^2\,E_F^{-1/2}\int_{-\infty}^{\infty}\frac{x^2e^x\,dx}{(e^x+1)^2}+\ldots\right]\quad\ldots(5)$$

Using standard integrals as

$$\int_{-\infty}^{\infty}\frac{e^x\,dx}{(e^x+1)^2}=1$$

$$\int_{-\infty}^{\infty}\frac{xe^x\,dx}{(e^x+1)^2}=0$$

and
$$\int_{-\infty}^{\infty}\frac{x^2e^x\,dx}{(e^x+1)^2}=\frac{\pi^2}{3}.$$

equation (5) becomes

$$I = \frac{2}{3}\left[E_F^{3/2} \cdot 2 \cdot \frac{1}{2} + 0 + \frac{3}{8}(kT)^2 E_F^{-1/2} \cdot 2 \cdot \frac{\pi^2}{3} + \dots \right]$$

$$\Rightarrow \quad I = \frac{2}{3} E_F^{3/2}\left[1 + \frac{\pi^2}{8}\left(\frac{kT}{E_F}\right)^2 + \dots \right] \qquad \dots(6)$$

Taking into account only upto second term of this integral as in (6) into eqn. (1), we have

$$N = \frac{4\pi V}{h^3}(2m)^{2/3}\frac{2}{3} E_F^{3/2}\left[1 + \frac{\pi^2}{8}\left(\frac{kT}{E_F}\right)^2 \right] \qquad \dots(7)$$

At absolute zero, $E_F = E_{F0}$

So eqn. (7) reduces to

$$N = \frac{4\pi V}{h^3}(2m)^{2/3} \times \frac{2}{3} E_{F0}^{3/2} \qquad \dots(8)$$

Now substituting equation (8) into (7), we have

$$E_{F0}^{3/2} = E_F^{3/2}\left[1 + \frac{\pi^2}{8}\left(\frac{kT}{E_{F0}}\right)^2 \right]$$

$$\Rightarrow \qquad E_F = E_{F0}\left[1 - \frac{\pi^2}{12}\left(\frac{kT}{E_{F0}}\right)^2 \right] \qquad \dots(9)$$

Eqn. (9) indicates that the Fermi energy is not constant but decreases slightly with an increase in temperature. However, the value of the factor $(kT/E_{F0})^2$ is very small at room temperature and the Fermi energy is considered to be a constant. Hence, the subscript '0' is dropped from eqn. (8).

9. WORK FUNCTION

Any metal contains a large no. of free electrons (typically electron concentration in a metal $\approx (10^{29}$ per meter^3). At normal temperature most electrons occupy the levels up to E_F. A conduction electron may move freely within the interior of the metal. But it can not escape from the metal. The electron has to surmount the potential barrier on the metal surface to escape from the metal. If an electron attempts to escape it will induce a positive charge on the surface of the metal. So unless the electron possesses enough energy to move away from the region of influence of the induced positive

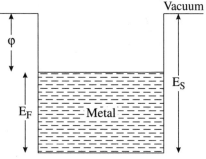

Fig. 13   PE diagram of metal showing work function.

charge it will be compelled to fall back the potential well. So energy is to be supplied to it to remove it from the metal. An electron at the lowest level has energy E_0 and an electron at Fermi level has energy $(E_0 - E_F)$ as in Fig. 13.

- The minimum energy required for an electron to be just emitted from the metal surface is called the Work Function (Φ) of that particular metal. Hence, $\Phi = E_0 - E_F$.

10. ELECTRON EMISSION

- The liberation of electrons from the metal surface is known as Electron Emission.

It is possible in the following four ways :

(a) Photoelectric Emission : Here metal is illuminated with a high frequency light.

(b) Field Emission : Here metal is subjected with a high electric field.

(c) Secondary Emission : Here metal is bombarded with a high energy electron.

(d) Thermoionic Emission : Here metal is heated to a high temperature, the electrons thus emitted are also called 'Thermions'.

11. THERMOIONIC EMISSION

When a metal is heated the electrons gain thermal energy (kT) and when it is more that Φ the electrons are emitted from the surface. The thermal energy, in excess of Φ will be converted into the KE of the electrons.

Requirements for Thermoionic Emitters : (The emitters are normally at zero potential or at negative potential with respect to an Anode $\Rightarrow$ Cathodes)

(a) Low work function.

(b) High melting point (m.p.), (m.p. > 1500°C) of the material.

(c) High mechanical strength. This is to withstand ion bombardment arises by the collisions between impurity gas molecules present (even in high vacuum) and energized electrons.

Materials of cathodes normally used :

1. Tungsten :

- $\Phi = 4.52$ eV (high);
- m.p. = 3300°C;
- high mechanical strength consistent emission.

Problem : Very high Φ, so its operating temperature 2300°C

2. Thoriated Tungsten : Tungsten filament coated with a layer of Thorium.

- $\Phi = 2.63$ eV (not high) operating temperature $> 1700°C$.
3. **Oxide coated cathode :** Commonly used.
- $\Phi = 1.1$ eV (not high) operating temperature $\approx 700 - 900°C$;
- A nickel ribbon or cylinder coated with a thin layer of Barium or strontium oxides.

12. RICHARDSON'S EQUATION

Richardson determined the emission current density of the emitted electrons. In the Free Electron model of the metal the electrons are free to move throughout the volume (V) of the metal. According to FD statistics the no. of electrons / volume having the momentum in the energy range E and $E + dE$ is

$$dn = n(E)\, dE = f(E)\, g(E)\, dE \qquad \text{...(a)}$$

We know, the DOS is given by,

$$g(E)\, dE = \frac{4\pi}{h^3} (2m)^{3/2} E^{1/2} . dE \qquad \text{...(b)}$$

However, $\qquad E = \dfrac{p^2}{2m} \Rightarrow dE = \dfrac{p}{m} dp$

Using these into eqn.(b) we have,

$$g(E)\, dE = \frac{4\pi}{h^3} . (2m)^{3/2} . \sqrt{\frac{p^2}{2m}} . \frac{p}{m} . dE = \frac{8\pi p^2 dp}{h^3} \qquad \text{...(c)}$$

Hence, as $\qquad f(E) = \dfrac{1}{e^{(E - E_F)/\kappa T} + 1}$

we have from eqn. (a) and (c),

$$dn = \frac{8\pi p^2 dp}{h^3}\, \frac{1}{e^{(E - E_F)/\kappa T} + 1} ;$$

This is shown in Fig. 14 below to be of a spherical shell of radius p and thickness dp.

Hence,

$$dn = \frac{2}{h^3}\, \frac{4\pi p^2 dp}{e^{(E - E_F)/\kappa T} + 1} \qquad \text{...(1)}$$

Now, p_x, p_y and p_z are the components of momentum of the electron along X, Y and Z axes such that $p^2 = p_x^2 + p_y^2 + p_z^2$.

Then the volume of the momentum space lying between p_x and $p_x + dp_x$, p_y and $p_y + dp_y$, d_z and $p_z + dp_z$ is

$$dp_x\, dp_y\, dp_z = 4\pi p^2 dp.$$

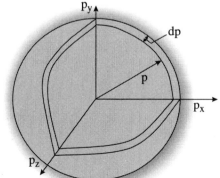

Fig. 14   A solid sphere of radius p in momentum space.

Hence, the no. of electrons per unit volume having momentum components lie between p_x and $p_x + dp_x$, p_y and $p_y + dp_y$, p_z and $p_z + dp_z$ is given by (using eqn. (1))

$$dn = \frac{2}{h^3} \frac{dp_x \, dp_y \, dp_z}{e^{(E - E_F)/\kappa T} + 1}$$

$$\Rightarrow \qquad dn = \frac{2}{h^3} \frac{d(mv_x) \, d(mv_y) \, d(mv_z)}{e^{(E - E_F)/\kappa T} + 1}$$

Here, v_x, v_y and v_z are the components of velocity of the electrons along X, Y and Z axes.

$$\Rightarrow \qquad dn = \frac{2m^3}{h^3} \frac{dv_x \, dv_y \, dv_z}{e^{(E - E_F)/\kappa T} + 1} \qquad \qquad ...(2)$$

Above equation gives the no. of electrons per unit volume having velocity components in the ranges between v_x and $v_x + dv_x$, v_y and $v_y + dv_y$, v_z and $v_z + dv_z$.

As $kT \approx 0.3$ eV, so, $(E - E_F) >> kT$; hence, from equation (2),

$$\Rightarrow \qquad dn = \frac{2m^3}{h^3} \frac{dv_x \, dv_y \, dv_z}{e^{(E - E_F)/\kappa T}}$$

$$\Rightarrow \qquad dn = \frac{2m^3}{h^3} e^{E_F/\kappa T} \, e^{-mv^2/2 \kappa T} \, dv_x \, dv_y \, dv_z$$

$$\Rightarrow \qquad dn = \frac{2m^3}{h^3} e^{E_F/\kappa T} \, e^{-\frac{1}{2 \kappa T} m (v_x^2 + v_y^2 + v_z^2)} \, dv_x \, dv_y \, dv_z$$

$$\Rightarrow \qquad dn = \frac{2m^3}{h^3} e^{E_F/\kappa T} \, e^{-\frac{1}{2} \frac{mv_x^2}{\kappa T}} dv_x \, e^{-\frac{1}{2} \frac{mv_y^2}{\kappa T}} dv_y \, e^{-\frac{1}{2} \frac{mv_z^2}{\kappa T}} dv_z \qquad ...(3)$$

Now we suppose that the surface of the metal lies in Y-Z plane and X-axis is normal to this surface, so the no. of electrons per unit volume in velocity range v_x and $v_x + dv_x$ is given by

$$\Rightarrow dn_x = \frac{2m^3}{h^3} e^{E_F/\kappa T} \, e^{-\frac{1}{2} \frac{mv_x^2}{\kappa T}} dv_x \int_{-\infty}^{\infty} e^{-\frac{1}{2} \frac{mv_y^2}{\kappa T}} dv_y \int_{-\infty}^{\infty} e^{-\frac{1}{2} \frac{mv_z^2}{\kappa T}} dv_z$$

$$\Rightarrow dn_x = \frac{2m^3}{h^3} e^{E_F/\kappa T} \, e^{-\frac{1}{2} \frac{mv_x^2}{\kappa T}} dv_x \left(\sqrt{\frac{2\pi\kappa T}{m}} \right) \left(\sqrt{\frac{2\pi\kappa T}{m}} \right)$$

$$\Rightarrow dn_x = \frac{4\pi m^2 \kappa T}{h^3} e^{E_F/\kappa T} \, e^{-\frac{1}{2} \frac{mv_x^2}{\kappa T}} dv_x \qquad \qquad ...(4)$$

When the metal is heated only those electrons can escape from the surface along X-axis which have

$$\frac{1}{2} mv_x^2 \geq E$$

$$\Rightarrow \qquad v_x^2 \geq \frac{2E}{m}.$$

$$\Rightarrow \qquad v_x \geq \sqrt{\frac{2E}{m}}$$

Hence the total number of electrons leaving unit area of the surface in unit time is

$$n_x = \int_{\sqrt{2E/m}}^{\infty} v_x \, dn_x$$

$$\Rightarrow \qquad n_x = \frac{4\pi m^2 \kappa T}{h^3} e^{E_F/\kappa T} \int_{\sqrt{2E/m}}^{\infty} e^{-\frac{1}{2}\frac{mv_x^2}{\kappa T}} v_x \, dv_x$$

we know,

$$\left[\int_{\sqrt{2E/m}}^{\infty} e^{-\frac{1}{2}\frac{mv_x^2}{\kappa T}} v_x \, dv_x = \frac{kT}{m} \cdot e^{-E/kT} \right]$$

Hence,
$$n_x = \frac{4\pi m \kappa^2 T^2}{h^3} e^{E_F/\kappa T} e^{-E/\kappa T}$$

$$\Rightarrow \qquad n_x = \frac{4\pi m \kappa^2 T^2}{h^3} e^{-(E - E_F)/\kappa T}$$

$$\Rightarrow \qquad n_x = \frac{4\pi m \kappa^2 T^2}{h^3} e^{-\phi/\kappa T} \qquad \qquad ...(5)$$

The emission current density 'J' is obtained by multiplying the no. of electrons leaving unit area of the surface in unit time with the electronic charge.

Hence,
$$J = n_x e = \frac{4\pi m e \kappa^2 T^2}{h^3} e^{-\phi/\kappa T}$$

$$\Rightarrow \qquad J = AT^2 e^{-\phi/\kappa T} \qquad \qquad ...(6)$$

Here, $A = \dfrac{4\pi m e \kappa^2}{h^3} \Rightarrow$ Constant, independent of the nature of the metal.

Equation (6) is the well-known Richardson's equation for thermionic emission.

Experimental verification : The emission current varies exponentially on work function and from equation (6) we have,

$$\frac{J}{T^2} = A \cdot e^{-\phi/\kappa T}$$

$$\Rightarrow \qquad \log\left(\frac{J}{T^2}\right) = \log A - \frac{\phi}{kT}$$

$$\Rightarrow \qquad \log\left(\frac{J}{T^2}\right) = \log A - \frac{\phi}{k}\left(\frac{1}{T}\right) \qquad \ldots(7)$$

From the eqn. (7) it is evident that the plot of $\log\left(\dfrac{J}{T^2}\right)$ versus $1/T$ should be a straight line with a slope $-\dfrac{\phi}{k}$ and an intercept $\log A$. This is shown in Fig. 15 aside and totally in agreement with the experimental results.

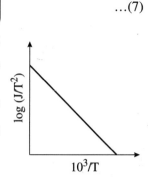

Fig. 15 Theoretical curve of $10^3/T$ vs $\log (J/T^2)$.

4

BAND THEORY

In early days of electronic industry vacuum tube devices were used for different electrical applications. Development of semiconductor devices (they often called solid state devices as they are used in solid form for applications) followed next and became popular due to their manifold advantages. The understanding of semiconductor materials requires different approach other than metals. The electrons in solid crystalline materials are under the influence of a periodic electric field produced by regularly placed ions at their lattice sites of the crystal. Kronig-Penney applied the Schrödinger wave equation to the electrons in such a condition and analyzed their behaviour. They explained the formation of forbidden band quantitatively, based on which different materials can be classified. Each band consisting of a sequence of closely spaced energy levels. The information about the energy band formation by putting restrictions on electron's motion inside the crystal is obtained from the band theory. It gives us the quantum mechanical visualization of transfer of electrons from one energy level to an empty energy level by surpassing the forbidden gap and thus explains the physical properties of metals, insulators and semiconductors. Without knowledge of band theory it is impossible to understand the principle and working of semiconductor devices.

1. INTRODUCTION

Solid contains of a large number of atoms. When we deal with a large no. of interacting particles the problem of calculating the electronic wave functions and energy levels is extremely complicated. In a two atom system the outer valence shell electron's energy level appear as single when they are far away from each other. As they come closer energy level starts to split into two energy level as shown in Figure 1(a). As the number of interacting atoms increases, the corresponding energy levels of electrons lie within those splitted energy region as shown in Figure 1(b). In solid these outer valence electronic energy level overlap each other and form energy band as shown in Figure 1(c). Based on energy band theory materials can be classified into conductor, semi-conductor and insulator.

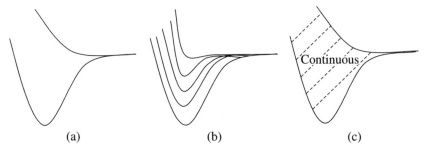

| (a) | (b) | (c) |

Fig. 1   Splitting of energy level (a) Two atoms come closer and at moderate distance
(b) Here number of interacting atoms increases and hence more splitting
(c) For a large number of interacting atoms, energy levels becomes continuous.

Let us consider the example of carbon and silicon. For Carbon, the electronic structure is $1s^2, 2s^2, 2p^2$ and for Silicon, the electronic structure is $1s^2, 2s^2, 2p^6, 3s^2, 3p^2$. Here the neighbouring outer valence electronic energy band interact each other, and forms conduction band & valence band separated by a forbidden band. The energy bands corresponding to $2s, 2p$ energy levels interact each other and form valence band in which all the energy levels are occupied by electrons while conduction band is completely empty. At equilibrium distance (r_0) the forbidden energy gap between valence band and conduction band is small such that electron can jump over this energy gap by getting energy from any external source. In semiconductor this energy gap is small, while in insulators this gap is large enough. In insulators electron from valence band can not get sufficient energy to cross the forbidden gap. In conductor conduction band overlap into valence band, so free electrons are always available in conduction band. Similar to Silicon,

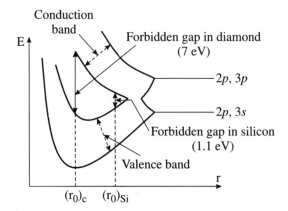

Fig. 1(d)   The energy band gap for diamond and silicon.

overlapping of energy band corresponding to 3*s*, 3*p* takes place in Carbon, and, a large energy gap occurs between valence band and conduction band at equilibrium. The energy band gap for different types of material is shown in Figure 1(d).

2. | THE KRONIG-PENNEY MODEL

In case of free electron theory (FET) it is assumed that the electron moves in a region of constant potential inside a one or three dimensional potential well. This theory fails to explain why some materials are good electrical conductor white others are not.

This variation of electrical conductivity became nonexistent in FET. The essential features of the behaviour of electrons in a periodic potential is discussed by Kronig and Penney in a one dimensional model. This model introduced the origin of forbidden band. According to this model electron moves in a periodically varying potential. The potential is minimum at positive ion site and maximum between the two ions. This periodic potential is shown in Figure 2(a). He has assumed that the PE of an electron has the form of periodic array of square wells separated by a distance as shown in Figure 2(b).

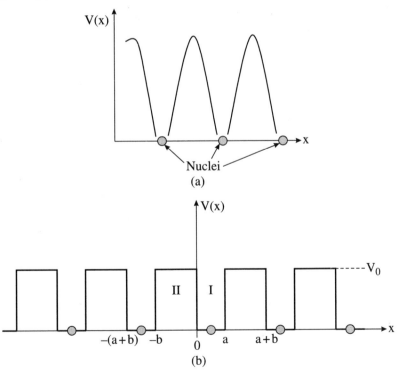

Fig. 2 (a) The motion of an electron under actual periodic potential in a crystal lattice (b) The motion of an electron under periodic potential as assumed by Kronig and Penney.

The periodic square well potential has the period $(a + b)$ and satisfy the following boundary conditions :

$$V(x) = 0 \text{ for } 0 < x < a ;$$
$$V(x) = V_0 \text{ for } -b < x < 0 ;$$

Each of the PE wells is a rough approximation for the potential in the vicinity of an atom.

Hence, the corresponding Schrödinger equations for the two regions I & II are of the form

$$\frac{d^2 \Psi_1(x)}{dx^2} + \frac{2m}{\hbar^2} E \Psi_1(x) = 0 \quad \text{for } 0 < x < a \quad \ldots(1)$$

$$\frac{d^2 \Psi_2(x)}{dx^2} + \frac{2m}{\hbar^2} (E - V_0) \Psi_2(x) = 0 \quad \text{for } -b < x < 0 \quad \ldots(2)$$

Here Ψ_1 & Ψ_2 are the wave function in the two regions I & II respectively. From the actual point of view, we have $E < V_0$.

So we define two real quantities α and β as

$$\alpha^2 = \frac{2mE}{\hbar^2}$$

and

$$\beta^2 = \frac{2m}{\hbar^2} (V_0 - E) \quad \ldots(3)$$

The general solutions of eqns. (1) & (2) are

$$\Psi_1 = A e^{i\alpha x} + B e^{-i\alpha x}$$

and

$$\Psi_2 = C e^{\beta x} + D e^{-\beta x}$$

Here A, B, C & D are constants in regions I and II. Their values can be obtaining on applying the continuous properties of the wave functions Ψ_1 & Ψ_2. So they must satisfy the following boundary conditions as

$$(\Psi_1)_{x=0} = (\Psi_2)_{x=0} \quad \text{and} \quad \left(\frac{d\Psi_1}{dx}\right)_{x=0} = \left(\frac{d\Psi_2}{dx}\right)_{x=0} \quad \ldots 4(a)$$

and

$$(\Psi_1)_{x=a} = (\Psi_2)_{x=-b} \quad \text{and} \quad \left(\frac{d\Psi_1}{dx}\right)_{x=a} = \left(\frac{d\Psi_2}{dx}\right)_{x=-b} \quad \ldots 4(b)$$

However, some mathematical correction factor is needed in equation 4(b) to make it consistent with the periodicity of the lattice. Since for a periodic lattice

$$V(x) = V(x + a + b)$$

it is expected that these wave functions will also exhibit the same periodicity. Bloch introduced a new function, named as Bloch function, which express this periodicity property as

$$\Psi(x) = u_k (x) . e^{ikx} \quad \ldots(5)$$

Here $u_k(x)$ is a potential function and is defined as

$$u_k(x) = u_k(x + a + b) \qquad \text{...(6)}$$

So equation (5) can be written as

$$\Psi(x + a + b) = u_k(x + a + b) \cdot e^{ik(x + a + b)} = u_k(x) \cdot e^{ikx} e^{ik(a + b)}$$

[using eqn. (6)]

$$\Rightarrow \qquad \Psi(x + a + b) = \Psi(x) \cdot e^{ik(a + b)}$$

$$\Rightarrow \qquad \Psi(x) = \Psi(x + a + b) \cdot e^{-ik(a + b)} \qquad \text{...(7)}$$

So, for $x = -b$, we have from eqn. (7),

$$\Psi_2(-b) = \Psi_1(a) \cdot e^{-jk(a + b)}$$

and

$$\left(\frac{d\Psi_2}{dx}\right)_{x = -b} = \left(\frac{d\Psi_1}{dx}\right)_{x = 0} \cdot e^{-ik(a + b)} \qquad \text{...(8)}$$

Here equation 4(b) is replaced by equation (8). So from the boundary conditions in equations 4(a) & (8), we have,

$$A + B = C + D \; ;$$

$$i\alpha(A - B) = \beta(C - D) \qquad \text{...(9)}$$

$$Ce^{-\beta b} + De^{\beta b} = e^{-ik(a + b)}[Ae^{i\alpha a} + Be^{-i\alpha a}]$$

$$\beta Ce^{-\beta b} - \beta De^{\beta b} = i\alpha \cdot e^{-ik(a + b)}[Ae^{i\alpha a} - Be^{-i\alpha a}] \qquad \text{...(10)}$$

Solution of equation (10) will have non-vanishing solution if and only if the determinant of the coefficients A, B, C & D vanish, i.e.,

$$\begin{vmatrix} 1 & 1 & -1 & -1 \\ i\alpha & -i\alpha & -\beta & \beta \\ e^{ik(a + b)} \cdot e^{i\alpha a} & e^{-ik(a + b)} \cdot e^{-i\alpha a} & -e^{-\beta b} & -e^{\beta b} \\ i\alpha e^{ik(\alpha + b)} \cdot e^{i\alpha a} & -i\alpha e^{-ik(a + b)} \cdot e^{-i\alpha a} & -\beta e^{-\beta b} & \beta e^{\beta b} \end{vmatrix} = 0$$

By solving we obtain the following condition,

$$\frac{\beta^2 - \alpha^2}{2\alpha\beta} \sinh(\beta b) \sin(\alpha a) + \cosh(\beta b) \cos(\alpha a) = \cos k(a + b) \quad \text{...(11)}$$

To obtain a more convenient equation Kronig and Penney consider the potential barrier to be a δ-function, i.e., V_0 tends to infinity when b approaches zero, but bV_0 remains finite.

Under these assumptions, when $V_0 \to \infty$, $\beta^2 \gg \alpha^2$.

$$\Rightarrow \qquad \frac{\beta^2 - \alpha^2}{2\alpha\beta} \approx \frac{\beta^2}{2\alpha\beta} = \frac{\beta}{2\alpha} \cdot$$

Also when $b \to 0$, $\sinh(\beta b) \to \beta b$, $\cosh(\beta b) \to 1$.

So,

$$\frac{\beta^2 - \alpha^2}{2\alpha\beta} \sinh(\beta b) = \frac{\beta}{2\alpha} \times \beta b$$

$$= \frac{b}{2\alpha} \times \frac{2m}{\hbar^2}(V_0 - E) \approx \frac{mV_0 b}{\hbar^2 \alpha} \qquad \text{(as } V_0 \gg E)$$

Hence, using this eqn. (11) becomes

$$\frac{mV_0b}{\hbar^2\alpha} \sin(\alpha a) + \cos(\alpha a) = \cos k\,a \quad (\text{as } b \to 0) \qquad \ldots(12)$$

Let,

$$P = \frac{mV_0ab}{\hbar^2}$$

then equation (12) becomes

$$P\frac{\sin(\alpha a)}{\alpha a} + \cos(\alpha a) = \cos k\,a \qquad \ldots(13)$$

Here, $P = \dfrac{ma}{\hbar^2}(V_0b)$ is a measure for the area 'V_0b' of the potential barrier (width = b and height = V_0). In other words 'P' physically means the binding of a given electron more strongly to a particular potential well. Eqn. (13) is represented schematically as $P\dfrac{\sin(\alpha a)}{\alpha a} + \cos(\alpha a)$ vs αa for $P = \dfrac{3\pi}{2}$ in Fig. 3.

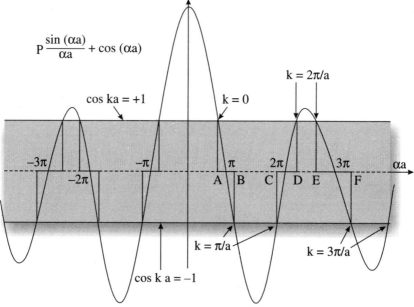

Fig. 3   Variation of $\left[P \cdot \dfrac{\sin(\alpha a)}{\alpha a} + \cos \alpha a\right]$ with (αa) for $P = 3\pi/2$ sharing allowed region for solution of eqn. (13).

Here, $\alpha^2 = \dfrac{2m}{\hbar^2}E$, so '$\alpha a$' represents a measure of energy as 'a' is constant. Also right hand side term is '$\cos k\,a$', so it only takes values between -1 to $+1$, as also shown in above Figure 3. So the condition of equation (13) can be satisfied only for values of αa for which left hand side lies between -1 and $+1$ (shaded region in the Figure 3 only). It is also clear that there are two different values of α for same 'k' value, while at $k = 0$ there is only one 'α'. From Figure 3 we see that the allowed limits of energy are as AB, CD, EF etc. These are thus the allowed energy bands. Also, $AB < CD < EF$ and $BC > DE$ and so on.

So we arrive at the following conclusions :

(i)   The energy spectrum of the electrons consists of a no. of allowed energy bands separated by forbidden regions (like BC, DE).

(ii) Width of the allowed bands increases with an increase of αa.

$\Rightarrow$ as $\dfrac{P}{\alpha a}$ sin αa decreases with an increase of αa,

Sharpness of the above curve reduces and becomes more and more flat as αa increases.

(iii) Width of the allowed band decreases with increase in P, *i.e.*, with more binding energy of electrons.

$\Rightarrow$ For $P \to \infty$, spectrum becomes a line.

Also, if $P \to \infty$, left side of equation (13) has to stay within ± 1,

so, $\dfrac{\sin \alpha a}{\alpha a} \to 0$, *i.e.*, $\sin \alpha a \to 0$

$\Rightarrow \qquad\qquad\qquad \alpha a = \pm\, n\pi$

$\Rightarrow \qquad\qquad\qquad \alpha^2 = \dfrac{n^2 \pi^2}{a^2} = \dfrac{8\pi^2 mE}{h^2}$

$\Rightarrow \qquad\qquad\qquad E = \dfrac{n^2 h^2}{8ma^2}$

This expression shows that the energy of the particle is discrete.

(iv) On the other hand, if $P = 0$, then equation (13) leads to

$$\cos \alpha a = \cos ka$$

$$\alpha = k \quad\Rightarrow\quad k^2 = \alpha^2 = \dfrac{8\pi^2 mE}{h^2}$$

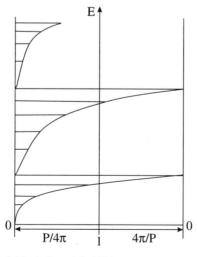

Fig. 4 The allowed (shaded) and forbidden energy ranges as a function of P.

Hence,
$$E = \frac{h^2 k^2}{8\pi^2 m}$$

So,
$$E = \frac{h^2 k^2}{8\pi^2 m} = \frac{h^2 (2\pi/\lambda)^2}{8\pi^2 m} = \frac{h^2}{2m\lambda^2}$$

Using de-Broglie's formula for wave-particle duality,
$$E = \frac{h^2}{2m}\left(\frac{p}{h}\right)^2 = \frac{p^2}{2m}.$$

This is just equivalent to the case of free particle.

The conclusions are summarized in the Figure 4 aside. Here we plotted energy as a function of P.

I   For $P = 0$ the free electron model and energy spectrum is (quasi) continuous.

II   For $P = \infty$, we have the line spectrum.

III   For any other P we get the energy spectrum by drawing a vertical line. The shaded area corresponds to allowed bands.

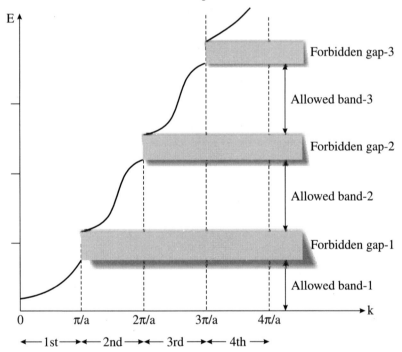

Fig. 5 The energy is a function of k for $P = 3\pi/2$ in one half showing different Brillouin zones.

The consequence of the limitations of values of 'cos ka' between -1 & $+1$, as shown in equation (13), is that only certain values of αa (and hence E) are allowed. From these values we can obtain a plot of E as a function of wave number k. This is shown in Figure 5 above. On the left side of the k (*i.e.*, for negative k) there is also a replica of the right side curve. Shaded region in the plot shows allowed energy values in between this shaded region there are disallowed energy values which do not satisfy equation (13). This band is known as forbidden band.

The discontinuities in E versus k curve occur for $\kappa = \dfrac{n\pi}{a}$, where $n = 0, \pm1, \pm2, \pm3, \ldots$ [as $\cos ka = \pm1$, so $\cos ka = \cos n\pi$].

These k-values define the boundary of 1st, 2nd, etc Brillouin Zones (BZ). So, 1st BZ lies between $+\pi/a$ to $-\pi/a$, 2nd BZ lies between $+\pi/a$ to $2\pi/a$ (1st part) and between $-\pi/a$ to $-2\pi/a$ (2nd part) and so on for 3rd, etc.

From the free electron model E versus k curve is parabolic as clear from the equation $E = \dfrac{\hbar^2 \kappa^2}{2m}$ and its curve is shown below in Figure 6(a). However, the actual E versus k curve from band theory (Kronig-Penney Model) is different as in Figure 6(b) below.

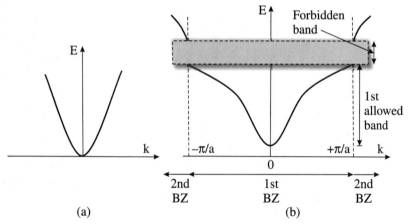

Fig. 6 (a) E vs k curve using free electron model
(b) E vs k curves using Kronig-Penney model in 1st BZ

The above Figure shows a discontinuity at $\pm\pi/a$ for 1st BZ and so on.

3. ENERGY LEVEL SPLITTING (ENERGY BAND)

The transformation of a single energy level into two or more separate levels is known as Energy Level Splitting. More atoms come together; more valence electron orbitals combine to form molecular orbital. The energy differences between them become smaller and smaller.

When two atoms come close one energy level splits into two (as shown in Figure 7(a)).

When three atoms come close one energy level splits into three (as shown in Figure 7(b)).

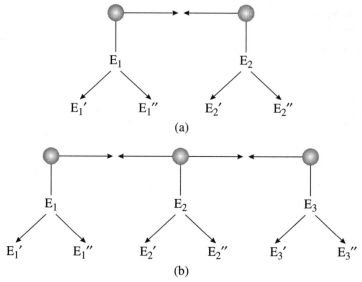

Fig. 7   (a) Energy level splitting for two interacting atoms
(b) Energy level splitting for three interacting atoms.

Hence, when N atoms come close the original level splits into N.

- The manifold levels resulting from splitting are so closely spaced that they form a virtual continuum. This is known as the Energy Band.

Alternate Approach (For Band Gap)

An electron wave propagating in a lattice at an angle of diffraction θ to the atomic plane gets diffracted at them in accordance to Bragg's Law, *i.e.*,

$$2d \sin \theta = n\lambda = n \frac{2\pi}{\kappa} \quad \Rightarrow \quad k = \frac{n\pi}{d \sin \theta}$$

According to above equation there is at least one series of the values of k corresponding to the integer n for which electrons are diffracted and do not pass freely into the crystal. So these values of k will be missing and hence as $E = \dfrac{\hbar^2 \kappa^2}{2m}$ is valid for free electron, also the energies corresponding to those values of k are not permitted for electrons in the crystal. Thus the electron energies are divided into the Forbidden bands and Allowed bands.

4. BRILLOUIN ZONES (BZ)

- BZ gives a zone having permissive values of k having allowed energy states.
- From both end points of this zone Forbidden band starts.

(a) For One Dimensional lattice : We know for the starting point of discontinuity in E-k curve, $\cos ka = \pm 1 \implies ka = \pm n\pi \implies k = \pm \dfrac{n\pi}{a}$, where, $n = 1, 2, 3, ...$

The region between the values of k at which 1st energy discontinuity occurs is called the 1st BZ, *i.e.*, when $k = \pm \pi/a$. Similarly, for 2nd BZ, $k = \pm 2\pi/a$ and is so on. All the BZs are shown in Figure below.

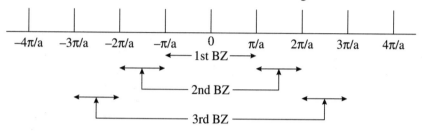

Fig. 8(a)   Brillouin zones for a one dimensional lattice.

(b) For Two Dimensional lattice : Using Bragg's diffraction law, we have,

$$k = \pm \frac{n\pi}{a \sin \theta} \qquad \qquad ...(1)$$

two dimensional square lattice, the reflection from horizontal rows of ions or atoms occur when $k_x = k \sin \theta$ (as in Figure 8(b)).

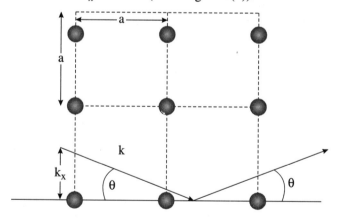

Fig. 8(b)   Bragg's reflection from a two dimensional lattice.

Hence, using equation (1),

$$k_x = n\pi/a \qquad \qquad ...(2)$$

Similarly, reflection from vertical rows occur when,

$$k_x = n\pi/a, \, k_y = n\pi/a \qquad \qquad ...(3)$$

So for first zone,

$$n = \pm 1 \quad \Rightarrow \quad k_x = \pm \pi/a \quad \text{and} \quad k_y = \pm \pi/a$$

Thus the first BZ is bounded along X and Y-axes by A, B, C and D as shown in Figure 8(c).

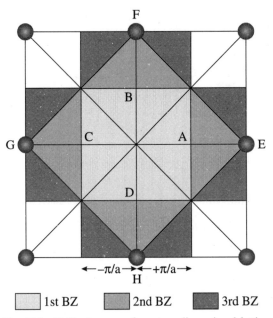

| 1st BZ | 2nd BZ | 3rd BZ |

Fig. 8(c)　Brillouin zones for a two dimensional lattice.

Now we consider a square lattice whose primitive translation vectors are $\vec{a} = a\hat{i}$ and $\vec{b} = a\hat{j}$, where a is the side of the square. So the corresponding reciprocal lattice translationals are

$$a^* = \frac{2\pi}{a}\hat{i} \quad \text{and} \quad b^* = \frac{2\pi}{a}\hat{j} \qquad \ldots(4)$$

So the Reciprocal lattice vector $\vec{G}$ can be written as

$$\vec{G} = n_1 a^* + n_2 b^* = \frac{2\pi}{a}(n_1\hat{i} + n_2\hat{j}) \qquad \ldots(5)$$

Now,
$$\vec{k} = k_x\hat{i} + k_y\hat{j} \qquad \ldots(6)$$

Hence using Bragg's condition,

$$2\vec{k} \cdot \vec{G} + G^2 = 0, \text{ we have,}$$

$$2k_x \frac{2\pi}{a} n_1 + 2k_y \frac{2\pi}{a} n_2 + \frac{4\pi^2}{a^2}(n_1 + n_2)^2 = 0$$

$$\Rightarrow \qquad n_1 k_x + n_2 k_y = \frac{\pi}{a}(n_1^2 + n_2^2) \text{ (only magnitude) } \ldots(7)$$

Here n_1 and n_2 are integers for diffraction by vertical and horizontal rows of ions.

For 1st zone, one integer = ±1 and other = 0

$\Rightarrow$ $\qquad\qquad\qquad k_x = \pm\,\pi/a\,(n_1 = \pm\,1,\, n_2 = 0)$

or $\quad\Rightarrow$ $\qquad\qquad\qquad k_x = \pm\,\pi/a\,(n_1 = 0,\, n_2 = \pm\,1)$

- When $k < \pi/a$, the electrons can move freely without being diffracted along X-axis.
- When $k = \pi/a$, the electrons are prevented along Y-axis.

The more k exceeds π/a more limited the possible direction of motion until when $k = \dfrac{\pi}{a}\sin 45° = \pi\sqrt{2}/a$, the electrons are diffracted even when they move diagonally (at 45° with k_x and k_y axes) inside the square. Here second zone begins, For this $n_1 = n_2 = \pm\,1$.

Hence, for $n_1 = +1$ and $n_2 = +1 \Rightarrow k_x = K_y = 2\pi/a$;

$\qquad\qquad\qquad\qquad$ for $n_1 = +1$ and $n_2 = -1 \Rightarrow k_x - k_y = 2\pi/a$

for $n_1 = -1$ and $n_2 = +1 \Rightarrow -k_x + k_y = 2\pi/a$;

$\qquad\qquad\qquad\qquad$ for $n_1 = -1$ and $n_2 = -1 \Rightarrow -k_x - k_y = 2\pi/a$

The above four equations describe a set of four lines at 45° to the k_x and k_y axes passing through E, F, G and H. Same thing occurs for 3rd BZ.

- For simple drawing we draw vectors from origin to all 8 lattice points. Then we draw planes which bisect these vectors perpendicularly. The smallest volume enclosed by the area between these planes is in the 1st BZ. Here it is a square of edge $2\pi/a$. The 2nd BZ is defined by the area between the smallest and next smallest area enclosed by the lines bisecting next nearest lattice vector to the origin.

---

5. | CONCEPT OF EFFECTIVE MASS AND HOLES (FROM *E-k* CURVES)

When the motion of electron is considered to be in a periodically varying potential, the mass of the electron changes. Under the application of electric field at some portion electron will accelerate with the electric field strength, while at some other portion it decelerates.

The mass of the electron under periodically varying electric field is known as the effective mass of the electron.

The effective mass, m^*, varies with the direction of motion of electron in the lattice. m^* can be large, small, even can also be negative depending on the position of electron inside the lattice.

Let the velocity of an electron is described by k. From the wave mechanical theory of particles, the particle velocity is equal to the group velocity of the waves representing the particle, *i.e.*, $v = \dfrac{d\omega}{dk}$, here, ω is the angular velocity of the de-Broglie waves.

Hence, the energy of the particle, $E = \hbar \omega$.

$$\Rightarrow \qquad\qquad v = \frac{d}{dk}(E/\hbar) = \frac{1}{\hbar}\frac{dE}{dk} \qquad\qquad \dots(1)$$

Also, using free electron model,

$$E = \frac{1}{2}mv^2 = \frac{1}{2}m\left(\frac{p}{m}\right)^2$$

$$\Rightarrow \qquad\qquad E = \frac{1}{2m}\left(\frac{h}{\lambda}\right)^2 = \frac{1}{2m}(\hbar\kappa)^2$$

Hence, $\qquad\qquad\qquad E \propto k^2$

However, in the Band theory, E is not proportional to k^2 and the **E–k curve** is shown in Figure 9(a).

The $v - k$ curve using equation (1) is also shown in Figure 9(b).

The following points are to be noted.

- At the top and bottom of the energy band $v = 0$, this is because at these points $\dfrac{dE}{dk} = 0$ (derivative of any curve at maximum or minimum position is always zero).
- The of velocity is maximum at $k = k_0$, which is totally different from free electron theory.

The Effective Mass

Now we see the case when an external electric field Σ is applied on an electron. We also assume that the BZ under consideration contains only one electron with charge 'e' so with an initial state k. When Σ is applied for a small time dt, the energy gained by it is,

$$dE = -e\Sigma v\, dt$$

[energy/work done = force × displacement

$$= -e\Sigma \times dx = -e\Sigma \times \frac{dx}{dt} \times dt = -e\Sigma v\, dt]$$

$$\Rightarrow \qquad\qquad dE = -e\Sigma\left(\frac{1}{\hbar}\frac{dE}{dk}\right)dt$$

$$\Rightarrow \qquad\qquad \frac{dk}{dt} = -\frac{e\Sigma}{\hbar} \qquad\qquad \dots(2)$$

The acceleration of the electron is

$$a = \frac{dv}{dt}$$

$$\Rightarrow \qquad\qquad a = \frac{1}{\hbar}\frac{d}{dt}\left(\frac{dE}{dk}\right)$$

$$\Rightarrow \qquad\qquad a = \frac{1}{\hbar}\frac{d}{dk}\left(\frac{dE}{dk}\right)\frac{dk}{dt}$$

$$\Rightarrow \qquad\qquad a = -\frac{1}{\hbar}\frac{d^2E}{dk^2}\left(\frac{e\Sigma}{\hbar}\right)$$

$$\Rightarrow \qquad\qquad a = -\frac{e\Sigma}{\hbar^2}\frac{d^2E}{dk^2} \qquad\qquad ...(3)$$

The force thus from eqn. (4) is

$$-e\Sigma = \frac{a}{(1/\hbar^2)\dfrac{d^2E}{dk^2}} \qquad\qquad ...(4)$$

Now comparing with the acceleration of a free electron of mass m we have, force = ma.

Hence, we observe that the electron behaves as if it had as effective mass m^* and is given by

$$m^* = \frac{1}{(1/\hbar^2)\dfrac{d^2E}{dk^2}} = \frac{\hbar^2}{\left[\dfrac{d^2E}{dk^2}\right]} \qquad\qquad ...(5)$$

Hence, the effective mass is totally dependent on $\dfrac{d^2E}{dk^2}$ and is shown in Figure 9(c). It indicates the importance of the E–k curve for the motion of electrons.

- $m^* - k$ curve shows that m^* is positive in the lower half of the energy band and negative in the upper half.
- At the inflection points m^* is infinite.

The electron response in the upper part of the first energy band to the applied field is very different from that of a free electron would do.

The effective mass concept arises due to the interaction of the electron wave packet with the periodic lattice. If the interaction with them is very large, i.e.,, there is a strong binding force between the electron and the lattice, it will be difficult for the electron to move; meaning that the electron has acquired a large (even infinite) effective mass. For negative effective mass let us consider an electron with k-value just less than π/a at the boundary. It will manage to move through the crystal. But suppose that a field is applied which should accelerate it and increase k. As the electron responds to the field it will meet the condition for Bragg reflection and will be

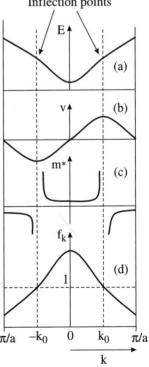

Fig. 9 The variation of E, v, m^* and f_k with k. The dotted vertical lines denote the inflection points.

scattered back in the opposite direction. In this way it will behave like a particle with a negative charge and negative mass.

When m^* is infinite the applied field causes no acceleration of the electron. The momentum gained by the electron due to the electric field is neutralized by the momentum lost by it to the lattice through reflection. Near the top of the band m^* is negative, the applied force here causes an acceleration in a direction opposite to that would normally be experienced by a free electron. Here the Bragg reflection is more. In this situation there is net decrease in the forward momentum of the electron. Thus the negative effective mass of an electron means electron responds to an applied electric field with a decrease in momentum. We may say that near the top of the band electron behaviour converts into hole behaviour.

At the bottom of the energy band $\dfrac{d^2E}{dk^2}$ rises with the values of 'k'. Thus m^* also start to increase. In this region electron behaviour is like that of free electron, where on increasing the applied electric field electron starts accelerating or the effective mass becomes positive here. Physical significance of effective mass of electron is that just by replacing the ordinary electronic mass by the effective mass of the electron every results of free electron theory of metals can easily be taken in account.

It is also convenient to use a factor 'f_k' which is a measure for the extent to which an electron in the state k is free. It is given by

$$f_k = \frac{m}{m^*} = \frac{m}{\hbar^2}\left(\frac{d^2E}{dk^2}\right) \qquad \qquad ...(6)$$

- If m^* is large, f_k is small and the electron behaves as a 'heavy' electron.
- If $f_k = 1$, the electron behaves as a 'free' electron.
- As in Figure 9(d), f_k = positive in lower half of the band and f_k = negative in upper half of the band.

6. CLASSIFICATION OF SOLIDS (BASED ON BAND THEORY)

The band theory leads to a distinction of solids into metals, insulators and semiconductors. We now consider a particular energy band which we assume to be filled with electrons up to $k = k_1$ as indicated in Figure 10 below.

The effective no. of free electrons in the band is given by

$$N_{eff} = \Sigma f_k \qquad \qquad ...(7)$$

Here summation extends all over the occupied states in the band. The no. of states in an interval 'dk' for a one dimensional lattice of length L is

$$dn = L\frac{dk}{2\pi} \qquad \qquad ...(8)$$

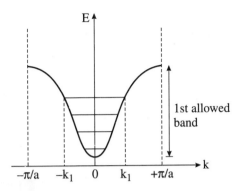

Fig. 10 *E–k* curve in 1st allowed band.

From eqn. (8), $N_{\text{eff}} = \int f_k \, dn$; also two electrons can occupy each of these states in the shaded region of the above figure. Hence,

$$N_{\text{eff}} = 2 \times \int f_k \, dn = 2 \frac{L}{2\pi} \int_{k=-k_1}^{+k_1} f_k \, dk \qquad \text{[using eqn. (8)]}$$

$$\Rightarrow \qquad N_{\text{eff}} = \frac{L}{\pi} \int_{k=-k_1}^{+k_1} \left(\frac{m}{\hbar^2} \frac{d^2E}{dk^2} \right) dk \qquad \text{[using eqn. (6)]}$$

$$\Rightarrow \qquad N_{\text{eff}} = \frac{mL}{\pi\hbar^2} \int_{k=-k_1}^{+k_1} \frac{d}{dk}\left(\frac{dE}{dk} \right) dk = \frac{mL}{\pi\hbar^2} \times 2 \times \int_0^{+k_1} d\left(\frac{dE}{dk} \right)$$

Hence,

$$N_{\text{eff}} = \frac{2mL}{\pi\hbar^2} \left(\frac{dE}{dk} \right)_{k=k_1} \qquad \qquad \ldots(9)$$

So, the effective no. of electrons in a completely filled band vanishes, as $\dfrac{dE}{dk} = 0$ at the top of the band. Also, N_{eff} is maximum for a band filled to the inflection point of the *E–k* curve as $\dfrac{dE}{dk}$ is maximum there.

Hence from the above discussions we understand that for a solid in which a certain no. of energy bands are completely filled, the other bands being completely empty, is an Insulator as in Figure 11(a) below.

In Figure 11(c) we have seen a solid which contains an energy band which is incompletely filled. It gives a Metal character.

From the above two contrasting nature of filled band we observe that Figure 11(c) situation can not occur only at $T = 0$, *i.e.*, when the crystal is in the lowest energy state. At $T > 0$ some electrons from the upper filled band will be excited into the next empty ('conduction') band and thus conduction becomes possible. If the forbidden gap (also called as 'Band gap, E_g') is of the order of several electron volts, the solid will remain as an insulator for all practical purposes. For example, diamond has $E_g \approx 7$ eV. It is an absolute insulator. For small E_g, *i.e.*, $E_g \approx 1$ eV, the no. of thermally excited electrons may become appreciable for conduction. In this case we designate the solid as Intrinsic semiconductor, as in Figure 11(b). Examples are Si, Ge, etc.

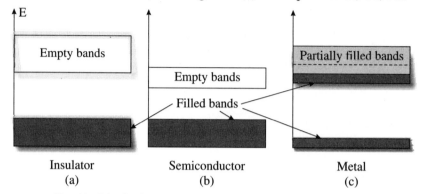

Fig. 11   Distribution of electrons at absolute zero for (a) insulator
(b) semiconductor (intrinsic) (c) metal the shaded regions are filled with electrons.

Metals

- The Fermi energy E_F is inside the conduction band
  e.g., in Na, E_F lies in the middle of the band
- At $T = 0$, all levels in the conduction band below E_F are filled with electrons while those above E_F are empty
- For $T > 0$, some electrons can be thermally excited to energy levels above E_F, but overall there is not much difference from the $T = 0$ case
- However electrons can be easily excited to levels above E_F by applying a (small) electric field to the metal
- Thus metals have high electrical conductivity

Insulators

- Here the Fermi energy E_F is at the midpoint between the valence band and the conduction band
- At $T = 0$, the valence band is filled and the conduction band is empty
- However the band gap energy E_g between the two is relatively large (~ 10 eV) (compared to $k_B T$ at room temperature *e.g.*)
- Thus there are very few electrons in the conduction band and the electrical conductivity is low

Semiconductors

- Again the Fermi energy E_F is midway between the valence band and the conduction band
- At $T = 0$, the valence band is filled and the conduction band is empty
- However for semiconductors the band gap energy is relatively small (1–2 eV) so appreciable numbers of electrons can be thermally excited into the conduction band
- Hence the electrical conductivity of semiconductors is poor at low T but increases rapidly with temperature

7. | THE CONCEPT OF A 'HOLE'

For intrinsic semiconductors at $T > 0$, a certain numbers of electrons may be excited thermally from the upper filled state into the conduction band. Thus some of the states in the normally filled band are unoccupied and creates 'hole' in this band.

We now consider a single 'hole' in the filled band of a one dimensional lattice and consider its influence on the collective behaviour of this band when an electric field is applied. If the electronic charge is '$-e$' and their velocities are v_i, the current associated with all electrons in the filled band in absence of an external field,

$$I = -e\, \Sigma\, v_i = -e\left[v_j + \underset{i \neq j}{\Sigma}\, v_i\right] = 0 \qquad \qquad ...(10)$$
$$\qquad i$$

$$\Rightarrow \qquad \qquad -e \underset{i \neq j}{\Sigma}\, v_i = ev_j \qquad \qquad ...(11)$$

Thus, if jth electron is missing and there we are having a hole, then the current without jth electron is

$$I' = -e \underset{i \neq j}{\Sigma}\, v_i \qquad \qquad ...(12)$$

From eqns. (11) and (12),

$$I' = ev_j \qquad \qquad ...(13)$$

Applying an external field F, the rate of change of this current I' due to the field is

$$\frac{dI'}{dt} = e\left(\frac{dv_j}{dt}\right)$$

$$\Rightarrow \qquad \frac{dI'}{dt} = e\left(\frac{1}{\hbar}\frac{d}{dt}\left(\frac{dE_j}{dk}\right)\right) \qquad \qquad [\text{as } v = \frac{1}{\hbar}\frac{dE}{dk}]$$

$$\Rightarrow \qquad \frac{dI'}{dt} = \frac{e}{\hbar}\frac{d}{dk}\left(\frac{dE_j}{dk}\right)\frac{dk}{dt}$$

$$\Rightarrow \qquad \frac{dI'}{dt} = -\frac{[e^2 F/\hbar]}{\hbar} \qquad [\text{from eqn. (2)}] \quad ...(14)$$
$$\overline{d^2 E_j/dk^2}$$

Since holes tend to reside in the upper part of a nearly filled band (as it is very easy for electrons of those part to leave and reach conduction band taking less energy from the surroundings), m_j^* is negative. In other word a band in which an electron is missing behaves as a positive hole with an effective mass $|m^*|$.

8. | INTRINSIC SEMICONDUCTORS

Carrier Concentration and Fermi Level

We have assumed here that the width of the allowed energy band is comparable to the forbidden gap of width E_g. Conduction band electrons have energy lying between E_C and ∞, and, Valence band electrons have energy lying between $-\infty$ to E_V. This is shown in Figure 12 aside.

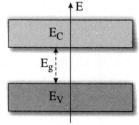

Fig. 12 Energy band diagram of an intrinsic semiconductor.

(a) Electron Concentration in Conduction Band (CB)

We know the total numbers of electrons per unit volume in conduction band is given by in terms of Fermi-Dirac distribution function $f(E)$ and density of states function $g(E)\,dE$ as

$$n_e = \int_{E_C}^{top} f(E)\,g(E)\,dE \qquad \qquad ...(1)$$

Here, $f(E)$ decreases as we rise through the CB because $f(E) = 0$ for $E \gg E_g$.

So we have set the upper limit for energy as infinity for CB.

Hence, eqn. (1) becomes,

$$n_e = \int_{E_C}^{\infty} f(E)\,g(E)\,dE \qquad \qquad ...(2)$$

Hence, putting the values of $f(E)$ and $g(E)\,dE$ we have,

$$n_e = \int_{E_C}^{\infty} \left[\frac{1}{e^{\{(E - E_F)/kT\}} + 1} \right] \times \left[\frac{4\pi}{h^3} (2m_e^*)^{3/2} \cdot (E - E_C)^{1/2} dE \right] \quad ...(3)$$

As $(E - E_F) \gg kT$, we can ignore the unit term in the $f(E)$ and so

$$n_e = \frac{4\pi}{h^3} (2m_e^*)^{3/2} \int_{E_C}^{\infty} [(E - E_C)^{1/2} \times e^{\{(E_F - E)/kT\}} dE]$$

$$\Rightarrow n_e = \frac{4\pi}{h^3} (2m_e^*)^{3/2} \int_{E_C}^{\infty} [\{(E - E_C)^{1/2} \times e^{\{(E_F - E_C)/kT\}} \times$$

$$e^{\{(E_C - E)/kT\}} dE] \quad ...(4)$$

Let, $\qquad \qquad \qquad y = (E - E_C)/kT$

$\Rightarrow \quad dE = kT\,dy$ and $[E - E_C]^{1/2} = [y]^{1/2} \cdot [kT]^{1/2}$

Putting these into eqn. (4) we have,

$$\Rightarrow n_e = \frac{4\pi}{h^3} (2m_e^*)^{3/2} (kT)^{1/2} \times e^{\{(E_F - E_C)/kT\}} \int_0^{\infty} [(y)^{1/2} \times e^{-y} \times (kT)\,dy]$$

$$...(4a)$$

We know,

$$\int_0^\infty [(y)^{1/2} \times e^{-y} \times dy] = \left(\frac{\pi}{4}\right)^{1/2}$$

Hence, from eqn. (4a),

$$\Rightarrow \quad n_e = \frac{4\pi}{h^3} (2m_e^*)^{3/2} (kT)^{1/2} \times e^{\{(E_F - E_C)/kT\}} \times \left(\frac{\pi}{4}\right)^{1/2} \qquad \text{...(4b)}$$

$$\Rightarrow \quad n_e = 2 \left(\frac{2\pi m_e^* kT}{h^2}\right)^{3/2} \times e^{\{(E_F - E_C)/kT\}} \qquad \text{...(5)}$$

This is the expression of the electron concentration in the CB.

(b) **Hole Concentration in Valence Band (VB)**

Here the probability for a state of energy E to be unoccupied is given by $[1 - f(E)]$.

Thus the total numbers of holes per unit volume is given by

$$n_h = \int_{bottom}^{E_C} [1 - f(E)] g(E) \, dE \qquad \text{...(6)}$$

As earlier we can change the lower limit as $-\infty$

Now, $\quad 1 - f(E) = 1 - \dfrac{1}{e^{\{(E - E_F)/kT\}} + 1} = \dfrac{e^{\{(E - E_F)/kT\}}}{e^{\{(E - E_F)/kT\}} + 1}$

However, for valence band, $E < E_F$, we have,

$$e^{\{(E - E_F)/kT\}} \ll 1$$

hence, $\qquad 1 - f(E) \approx e^{\{(E - E_F)/kT\}} \qquad \text{...(7)}$

Thus in VB, $[1 - f(E)]$ decreases exponentially as we move downwards from the top of the VB. Now, using eqn. (7) into eqn. (6) we have,

$$n_h = \int_{-\infty}^{E_v} e^{\{(E - E_F)/kT\}} \times \left[\frac{4\pi}{h^3} (2m_h^*)^{3/2} \cdot (E_V - E)^{1/2} \, dE\right] \qquad \text{...(8)}$$

On solving as in above eqns. (4a), (4b), we have,

$$n_h = 2 \left(\frac{2\pi m_h^* kT}{h^2}\right)^{3/2} \times e^{\{(E_V - E_F)/kT\}} \qquad \text{...(9)}$$

(c) **Value of Fermi Energy**

In the intrinsic semiconductor, $n_e = n_h$.

So equating eqns. (5) and (9) we have,

$$\Rightarrow \quad 2 \left(\frac{2\pi m_e^* kT}{h^2}\right)^{3/2} \times e^{\{(E_F - E_C)/kT\}} = 2 \left(\frac{2\pi m_h^* kT}{h^2}\right)^{3/2} \times e^{\{(E_V - E_F)/kT\}}$$

$$\Rightarrow \quad (m_e^*)^{3/2} \times e^{\{(E_F - E_C)/kT\}} = (m_h^*)^{3/2} \times e^{\{(E_V - E_F)/kT\}}$$

$$\Rightarrow \quad e^{\{(2E_F - E_C - E_V)/kT\}} = \left(\frac{m_h^*}{m_e^*}\right)^{3/2} \qquad \text{...(9a)}$$

$$\Rightarrow \quad \left\{ \frac{2E_F - E_C - E_V}{kT} \right\} = \frac{3}{2} \log \left(\frac{m_h^*}{m_e^*} \right)$$

$$\Rightarrow \quad E_F = \frac{E_C + E_V}{2} + \frac{3}{4} \log \left(\frac{m_h^*}{m_e^*} \right) \qquad \ldots(10)$$

For $\quad m_e^* = m_h^*, \log \left(\frac{m_h^*}{m_e^*} \right) = \log (1) = 0$

$$\Rightarrow \quad E_F = \frac{E_C + E_V}{2} \qquad \ldots(11)$$

So Fermi level lies exactly in the middle of the top of the VB and the bottom of the CB.

Now, putting the values of E_F into eqns. (5) and (9) separately we have,

$$\Rightarrow n_e = 2 \left(\frac{2\pi m_e^* kT}{h^2} \right)^{3/2} \times e^{\{(E_F - E_C)/kT\}} = 2 \left(\frac{2\pi kT}{h^2} \right)^{3/2} \times [(m_e^*)^2]^{3/4}$$
$$e^{\{(E_C + E_F)/2 - E_C/kT\}}$$

From eqn. (9a),

$$\Rightarrow n_e = 2 \left(\frac{2\pi kT}{h^2} \right)^{3/2} \times [(m_e^*) (m_h^*)]^{3/4} e^{\{(E_V - E_C)/2kT\}}$$

$$= 2 \left(\frac{2\pi kT}{h^2} \right)^{3/2} [m_e^* m_h^*]^{3/4} e^{\{-E_g/2kT\}} \qquad \ldots(12)$$

and $\quad n_h = 2 \left(\frac{2\pi kT}{h^2} \right)^{3/2} [m_e^* m_h^*]^{3/4} e^{\{-E_g/2kT\}} \qquad \ldots(13)$

Hence, $n_h = n_e = n_i$, the intrinsic carrier concentration of the semiconductor.

9. EXTRINSIC SEMICONDUCTORS

(a) **n-Type semiconductor** : For n-type semiconductors a donor level E_d is formed below the Fermi level. At a temperature T the density of conduction electrons from eqn. (5) is given by

$$n_e = 2 \left(\frac{2\pi m_e^* kT}{h^2} \right)^{3/2} e^{\{(E_F - E_C)/kT\}} \qquad \ldots(5)$$

The density of vacancies in the donor levels of energy E_d is given by

$$N_d^+ = n_d [1 - f(E)] = n_d \left[1 - \frac{1}{e^{\{(E_d - E_F)/kT\}} + 1} \right] = n_d \left[\frac{e^{\{(E_d - E_F)/kT\}}}{e^{\{(E_d - E_F)/kT\}} + 1} \right]$$

Here N_d^+ is concentration fo ionised donor atoms S. Assuming, the position a little above the donor level by a few 'kT' the above expression can be approximated as

$$N_d^+ = n_d\,[e^{\{(E_d - E_F)/kT\}}] \qquad \ldots(14)$$

However, the electron concentration in the CB and the vacancy concentration in the donor level should be equal. Hence, equating eqns. (5) and (14), we have,

$$n_e = 2\left(\frac{2\pi\, m_e^*\, kT}{h^2}\right)^{3/2} e^{\{(E_F - E_C)/kT\}} = n_d\left[e^{\{(E_d - E_F)/kT\}}\right]$$

$$\Rightarrow\ \log\left[2\left(\frac{2\pi\, m_e^*\, kT}{h^2}\right)^{3/2}\right] + \left\{\frac{E_F - E_C}{kT}\right\} = \log\,(n_d) + \left\{\frac{E_d - E_F}{kT}\right\}$$

$$\Rightarrow\qquad \left\{\frac{2E_F - E_C - E_d}{kT}\right\} = \log\frac{n_d}{\left[2\left(\dfrac{2\pi\, m_e^*\, kT}{h^2}\right)^{3/2}\right]}$$

$$\Rightarrow\qquad E_F = \left\{\frac{E_C + E_d}{2}\right\} + \frac{kT}{2}\,\log\frac{n_d}{\left[2\left(\dfrac{2\pi\, m_e^*\, kT}{h^2}\right)^{3/2}\right]} \qquad \ldots(15)$$

At absolute zero temperature, $E_F = \left\{\dfrac{E_C + E_d}{2}\right\} \Rightarrow$ the Fermi level lies at half way between donor level and the bottom of the CB.

With an increase in T, E_F decreases and at room temperature it comes below the donor level.

Substituting the value of E_F from eqn. (15) into eqn. (4) we have the free electron concentration in the CB as

$$n_e = 2\left(\frac{2\pi\, m_e^*\, kT}{h^2}\right)^{3/2} \exp\left\{\frac{E_F - E_C}{kT}\right\}$$

$$= 2\left(\frac{2\pi\, m_e^*\, kT}{h^2}\right)^{3/2} \exp\left[\frac{\left\{\dfrac{E_C + E_d}{2}\right\} + \dfrac{kT}{2}\,\log\dfrac{n_d}{\left[2\left(\dfrac{2\pi\, m_e^*\, kT}{h^2}\right)^{3/2}\right]} - E_C}{kT}\right]$$

$$\Rightarrow n_e = 2\left(\frac{2\pi\, m_e^*\, kT}{h^2}\right)^{3/2} \exp\left[\left\{\frac{E_d - E_C}{2kT}\right\} + \frac{1}{2}\,\log\frac{n_d}{\left[2\left(\dfrac{2\pi\, m_e^*\, kT}{h^2}\right)^{3/2}\right]}\right]$$

$$\Rightarrow n_e = 2\times\left(\frac{2\pi\, m_e^*\, kT}{h^2}\right)^{3/2} e^{\{(E_d - E_C)/2kT\}} \times \sqrt{\frac{n_d}{\left[2\left(\dfrac{2\pi\, m_e^*\, kT}{h^2}\right)^{3/2}\right]}}$$

$$\Rightarrow \qquad n_e = (2n_d)^{1/2} \left(\frac{2\pi \, m_e^* \, kT}{h^2} \right)^{3/4} e^{\{(E_d - E_C)/2kT\}}$$

The ionization energy is given by $\Delta E = E_C - E_d$ of the donors, then, we have,

$$\Rightarrow \qquad n_e = (2n_d)^{1/2} \left(\frac{2\pi \, m_e^* \, kT}{h^2} \right) e^{\{-\Delta E/2kT\}} \qquad \ldots(16)$$

From eqn. (16) we say that electron concentration is proportional to the square root of the donor concentration of the semiconductor. With more increase E_F reaches up to the middle of the CB and VB to make the material intrinsic.

(b) p-Type semiconductor : For p-type semiconductors a donor level E_a is formed above the Fermi level and the VB. At a temperature T the density of holes is similar to eqn. (5) and is given by

$$n_h = 2 \left(\frac{2\pi \, m_h^* \, kT}{h^2} \right)^{3/2} e^{\{(E_V - E_F)/kT\}} \qquad \ldots(17)$$

The density of electrons in the acceptor levels of energy E_a is given by

$$N_a^- = n_a \, [f\,(E)] = n_a \left[\frac{1}{e^{\{(E_a - E_F)/kT\}} + 1} \right] \qquad \ldots(18)$$

Here, N_a^- is ionized acceptor atoms concentration.

However, the hole concentration in the VB and the electron concentration in the acceptor level should be equal.

Also, assuming $E_a - E_F \gg a$ few kT, equating eqns. (17) and (18), we have,

$$2 \left(\frac{2\pi \, m_h^* \, kT}{h^2} \right)^{3/2} e^{\{(E_V - E_F)/kT\}} = n_a \left[\frac{1}{e^{\{E_a - E_F/kT\}}} \right]$$

On simplification,

$$\Rightarrow \qquad E_F = \left\{ \frac{E_a + E_V}{2} \right\} - \frac{kT}{2} \log \frac{n_a}{\left[2 \left(\frac{2\pi \, m_h^* \, kT}{h^2} \right)^{3/2} \right]} \qquad \ldots(19)$$

At absolute zero temperature,

$$E_F = \left\{ \frac{E_a + E_V}{2} \right\}$$

$\Rightarrow$ the Fermi level lies at half way between acceptor level and the top of the VB.

With an increase in T, E_F increases and at room temperature it comes above the acceptor level slightly. Hence, we have,

$$\Rightarrow \qquad n_h = (2nd)^{1/2} \left(\frac{2\,\pi\,m_h^*\,kT}{h^2} \right) e^{(-\Delta E/2kT)} \qquad \qquad \dots(20)$$

where $\Delta E = E_a - E_V$.

10. HALL EFFECT

If a metal or a semiconductor carrying a current is placed in a transverse magnetic field, a potential difference is produced in the direction normal to both the current and magnetic field direction.

Hall Effect : In the Figure 13 above electric field is applied along X-direction, magnetic field is in Z-direction. Hence, a potential difference due to Hall effect is developed in the rectangular sample along Y-direction.

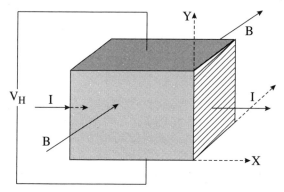

Fig. 13 An rectangular sample under Hall effect.

We now consider a rectangular plate of a p-type semiconductor. When a potential difference is applied for the application of electric field at the ends of it a current 'I' flows through it along X-direction as shown in next Figure 14. Hence the current due to charge carrier 'holes' in p-type semiconductor is given by,

$$I = peA \qquad \qquad \dots(1)$$

Here, p is concentration of holes in atoms/cc, A is the cross sectional area of the end face, e is the charge of a hole.

The current density is

$$J_x = \frac{I}{A} = pev_d \qquad \qquad \dots(2)$$

Here v_d is the average drift velocity of holes. By applying the external electric field now the potential difference between the front and rear faces F and F' is zero, as this plane along FF' is perpendicular to the plane of electric field. However, on application of B normal to the crystal surface and to the current flow, a transverse potential difference is produced between F and F'. This is called the Hall Voltage, V_H.

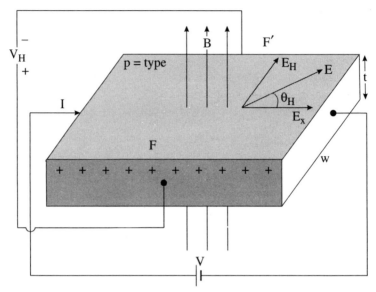

Fig. 14   Development of Hall voltage under crossed electric and magnetic field in a
rectangular sample.

Origin of V_H : Prior to the application of 'B' for p-type, holes move
in an orderly way parallel to the faces F and F'. On application of B they
experience a sideway deflection due to Lorentz force F_M as shown in Figure
below. The magnitude of magnetic force is

$$F_M = Bev_d \qquad \qquad ...(3)$$

Due to this force the holes are deflected towards the front surface F
and pile up there. Initially the material is electrically neutral everywhere.

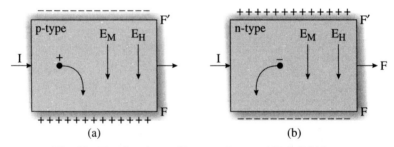

Fig. 15   The directions of Lorentz force and Hall field in an
(a) p-type and (b) n-type semiconductor.

However, as holes pile up towards the front side, corresponding negative
charges are left at rear face F'. As a result an electric field E_H is set up
between F and F'. This field starts opposing further piling up of holes and
thus an equilibrium results. Then F_E (due to 'I') balances F_M. The transverse
electric field, E_H, thus produced is known as the **Hall Field**. Equilibrium
state generally attained in $\approx 10^{-4}$ sec and after that holes flow once again
along X-direction parallel to the faces F and F'. Under equilibrium,

$$F_E = F_M \quad \text{and} \quad F_E = eE_H = e\,\frac{V_H}{w} \qquad \qquad ...(4)$$

Here w is the width of the semiconductor plate.
We know,

$$J_x = pev_d \quad \Rightarrow \quad v_d = \frac{J_x}{pe} \qquad \qquad ...(5)$$

From equation (3),

$$F_M = Bev_d = Be \times \frac{J_x}{pe} = \frac{BJ_x}{p} \qquad \qquad ...(6)$$

So at equilibrium, $\qquad F_e = F_M$

$$\Rightarrow \qquad\qquad e\,\frac{V_H}{w} = \frac{BJ_x}{p}$$

$$\Rightarrow \qquad\qquad V_H = \frac{wB}{pe}\,J_x = \frac{wB}{pe}\,\frac{I}{A} \qquad \qquad ...(7)$$

If 't' be the thickness of the plate (as in Figure 14), then, $A = wt$.

Hence, $\qquad\qquad V_H = \frac{wBI}{pewt} = \frac{BI}{pet} \qquad \qquad ...(8)$

Hall Coefficient :

- The Hall Coefficient, R_H, is defined as the Hall Field per unit magnetic induction per unit current density.

Hence, $\qquad\qquad R_H = \frac{E_H}{BJ_x}$

$$\Rightarrow \qquad\qquad R_H = \frac{V_H/w}{BJ_x} = \frac{BI/pet}{wB\,(I/wt)}$$

$$\Rightarrow \qquad\qquad R_H = \frac{1}{pe} \qquad \qquad ...(9)$$

From eqns. (8) and (9),

$$V_H = R_H\,\frac{BI}{t}$$

$$\Rightarrow \qquad\qquad R_H = \frac{tV_H}{BI} \qquad \qquad ...(10)$$

When the directions of B and I are like the directions as shown in Figure before, the sign of V_H is positive. However, for an n-type semiconductor V_H is negative for same directions of B and I. So by knowing the polarity of Hall voltage the type of a semiconductor can be known. Knowing the value of R_H (from eqn. (10)), the concentration of charge (majority) carriers can be determined (using eqn. (9)).

Hence, using Hall Effect we can

- determine the sign of charge carriers,

- determine the charge carrier concentration, and
- determine the mobility of charge carriers if conductivity of the material is known.

Determination of the Mobility of Charge Carriers

The net electric field 'E' is a vector sum of E_x and E_H (as in Figure 14). It acts an angle θ_H to the X-axis and is known as the **Hall Angle**. Here from the same Figure 14.

$$\tan \theta_H = \frac{E_H}{E_x}$$

$$\Rightarrow \quad \tan \theta_H = \frac{V_H/w}{\rho J_x}$$

$$\Rightarrow \quad \tan \theta_H = \frac{BJ_x/pe}{\rho J_x}$$

$$\Rightarrow \quad \tan \theta_H = \frac{B}{pe\rho} \qquad \text{(from eqn. (7))} \quad ...(11)$$

$$\Rightarrow \quad \tan \theta_H = R_H \frac{B}{\rho} = R_H B\sigma = \sigma R_H B = \mu_H B$$

$$\Rightarrow \quad \theta_H = \tan^{-1}(\mu_H B) \qquad ...(12)$$

Here, $\qquad \mu_H = \sigma R_H \Rightarrow$ Mobility of Holes.

- The Hall effect measurements in metals over a wide range of temperature established that R_H here does not depend on temperature, *i.e.*, free carrier concentration is independent of temperature. This is absolutely true.
- For semiconductors R_H drops sharply with a rise in temperature indicating that the concentration of free electrons increase with temperature.

5

PHOTOCONDUCTIVITY AND PHOTOVOLTAICS

The development of modern day instrumentation regarding any sorts of industries requires sensors, especially photosensors. The popularity of photosensors arises due to their compactness, low cost, simple operational procedures, high sensitivity, minimum noise and outstanding reliability besides other advantages. Any photosensor is basically developed on the basis of the interaction of light (or photons) with the structure of the materials. Photodetectors are basically semiconductor devices. They sense the optical signals through electronic process. The development of Laser as an optical source is very advantageous as a source of the optical signal. The broad optical band for ultraviolet to far-infrared used for optical communications requires high speed and very sensitive photodetectors. In any semiconductor based detectors the carriers are first created (known as the photogenerated carriers) inside the material on its exposure to incident light of proper frequency. These carriers then move around or multiplied by externally controlled processes and finally collected at the ends of the material to form an output electrical signal. This signal is totally dependent on the nature of the incident light, properties of the detecting material and the controlling options. Photodetectors are very important for fiber optic communication systems specially operated in near-infrared (0.8–1.6 µm) region.

Here we are studying two different photosensors, photoconductors and photovoltaic cells. First one is used for detection applications and second one for direct electrical power generation.

1. PHOTOCONDUCTIVITY IN INSULATING CRYSTALS

- Phenomenon of an increase in the electrical conductivity of an insulating crystal under the exposure of light radiation is known as **Photoconductivity**.

When a crystal is exposed to the light with the energy of the incident photons is higher than that of the band gap of the crystal, E_g, the free electron - hole pairs (EHP) are produced. These free electrons and holes serve as the carriers of conduction. The incident photons excite the electrons from the valence band (VB) into the conduction band (CB) and they become mobile. This photoconduction is of intrinsic type. Impurities in a crystal and imperfections of a crystal have also some contribution towards it. If the impurities are present, then even the photons having energy below the threshold for the production of EHP may be able to produce electrons and holes. Imperfections introduce discrete energy levels in the forbidden region and are often called 'Traps'. Decay of photoconductivity after the removal of the incident photons is natural. It is caused by the recombination of electrons and holes with each other.

2. VARIATION OF PHOTOCONDUCTIVITY WITH ILLUMINATION

We now consider here a simple model of a photoconductor as shown in Fig. (1) given below. It is also assumed that EHPs are uniformly produced throughout the volume of the crystal by irradiation with an external light source, the recombination occurs by direct annihilation of electrons with holes and that electrons leaving the crystal at one electrode are replaced by the electrons flowing in from opposite electrode. It is convenient to neglect the mobility of holes (≈ 20 cm²/V-s) in comparison to electrons (≈ 35 cm²/V-s). This model is a hypothetical one, rarely experienced practically, however, it gives a way for a correct model.

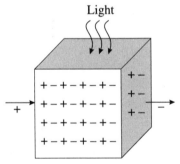

Fig. 1 A photoconductor illuminated with light.

The rate of change of electron concentration, n, is given by

$$\frac{dn}{dt} = L - Anp$$

$\Rightarrow$ $$\frac{dn}{dt} = L - An^2 \qquad \text{(using } n = p) \quad ...(1)$$

Here, L is the no. of photons absorbed per unit volume per unit time, and 'Anp' is the recombination rate [we know that the recombination rate is proportional to the product 'np', A is the proportionality constant].

We know that drift velocity of the carriers, v, is proportional to applied electric field, E, *i.e.*,

$$v \propto E \Rightarrow v = \mu E$$

Here, μ is the constant of proportionality and is known as the mobility of the medium. In steady state, $\dfrac{dn}{dt} = 0$ and $n = n_0$.

Hence using equation (1),

$$L = An_0^2$$

$\Rightarrow \qquad\qquad n_0 = \sqrt{L/A}$

$\Rightarrow \qquad\qquad n_0 \propto L^{0.5}$

and $\qquad\qquad \sigma = n_0 \, e\mu = e\mu \, (L/A)^{0.5}$ ...(2)

Equation (2) predicts the variation of photoconductivity, σ, with light level L as $L^{0.5}$. However, practically the actual exponent ranges between 0.5 to 1.0 or more as shown in Fig. 2 given next.

Decay : The decay of photoelectrons after switching off the light suddenly is

$$\frac{dn}{dt} = 0 - An^2 \qquad ...(3)$$

$\Rightarrow \qquad \displaystyle\int_{n_0}^{n} \frac{dn}{n^2} = -A \int_0^t dt$

Solution of Eqn. (3) is,

$$\frac{1}{n} = At + \frac{1}{n_0} \qquad ...(4)$$

where n_0 is the carrier concentration at $t = 0$ when light is just switched off.

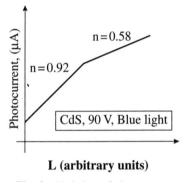

Fig. 2 Variation of photocurrent of cadmium sulphide with different intensity at 70 V bias voltage with blue light.

- **Response time, t_0 :** This is the time during which carrier concentration reduces to half of starting carrier concentration.

Hence, at $t = t_0$, $n = n_0/2$,

Using this in Eqn. (4),

$$\frac{1}{n_0/2} = At_0 + \frac{1}{n_0}$$

$\Rightarrow \qquad\qquad \dfrac{1}{n_0} = At_0$

$\Rightarrow \qquad\qquad t_0 = \dfrac{1}{An_0} = \dfrac{1}{A\sqrt{L/A}}$

$\Rightarrow \qquad\qquad t_0 = \sqrt{\dfrac{1}{LA}}$

$\Rightarrow \qquad\qquad t_0 = \dfrac{1}{L}\dfrac{L}{\sqrt{LA}}$

$\Rightarrow \qquad\qquad t_0 = \dfrac{1}{L}\sqrt{\dfrac{L}{A}} = \dfrac{n_0}{L}$

$\Rightarrow$ $$t_0 = \frac{n_0}{L} \qquad \qquad \ldots(5)$$

So, $$t_0 = \frac{1}{L}\left(\frac{\sigma}{e\mu}\right) = \frac{\sigma}{e\mu L} \qquad \ldots(6)$$

- $\therefore$ $t_0 \propto \sigma \Rightarrow$ Response time is directly proportional to σ at a given light intensity level.
- Good photoconductors have large t_0. These are rarely observed in practice.

Sensitivity or Gain Factor (G)

This can be calculated as the ratio of the numbers of carriers crossing the unit area of specimen to the numbers of photons absorbed by that area in the specimen. It is represented by 'G'.

So for a specimen of thickness, 'd' and the cross-section area unity we have

$$G = \frac{\text{Particle Flux}}{L \times (d \times 1)} \qquad \ldots(7)$$

We know, flux is the rate of flow of carriers per unit area per unit time and is given by 'F_n'.

Also, current density,

$$J_n = \sigma E = n_0 e\mu \, (V/d)$$

A potential V produces the particle flux F_n is given by

$$F_n = \frac{\mu V}{d} n_0$$

$\Rightarrow$ $$F_n = \frac{\mu V}{d} (Lt_0)$$

$\Rightarrow$ $$F_n = \frac{\mu V}{d} \times L \times \frac{1}{\sqrt{LA}}$$

Hence, $$F_n = \frac{\mu V}{d} \times L \times \frac{1}{\sqrt{LA}} \qquad \ldots(8)$$

So, from eqn. (7), $$G = \frac{F_n}{Ld}$$

$\Rightarrow$ $$G = \frac{1}{Ld} \times \frac{\mu V}{d} \times L \times \frac{1}{\sqrt{LA}}$$

$\Rightarrow$ $$G = \frac{\mu V}{d^2 \, (LA)^{1/2}} \qquad \ldots(9)$$

Now, let T_e is the lifetime of the electron before illumination and it is nothing but the response time, t_0, so,

$$T_e = t_0 = \frac{1}{\sqrt{LA}}$$

Let T_d is the transit time of an electron between the electrodes and is given by

$$T_d = \frac{d}{v_d} = \frac{d}{\mu E}$$

$$\Rightarrow \qquad T_d = \frac{d}{\mu V/d} = \frac{d^2}{\mu V}$$

So, from equation (9),

$$G = \frac{\mu V}{d^2 (LA)^{1/2}} = \frac{T_e}{T_d} \qquad \qquad ...(10)$$

The expression of G as given in equation (10) is quite general. However, this theoretical value of t_0 and the experimental value of t_0 do not agree and in some instances the discrepancy is $\approx 10^8$ times. The result then indicates :

- Model is a failure.
- We should look out for some missing phenomena not considered in this expression.

3. EFFECT OF TRAPS

- A trap is an energy level in the forbidden gap of the crystal which is capable of capturing a hole or electron. This captured carrier may be reemitted later on and may move to another trap.

Two Types of Traps

1. Recombination Centre : Here electrons and holes recombine and thermal equilibrium result.

2. Trap Centre : Here no recombination occurs. However, it affects the freedom of motion of charge carriers.

All these are shown in Fig. 3 given below. We now consider a crystal

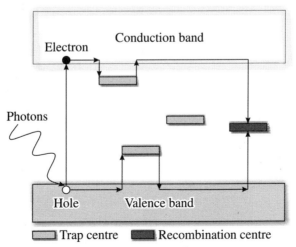

Fig. 3   Trap and recombination centres in energy band picture of a semiconductor.

with N electron trap levels per unit volume. By considering the temperature to be sufficiently low so that concentration of thermal carriers may be neglected, we can also write

$$\frac{dn}{dt} = L - An\,(n + N) \qquad \qquad ...(1)$$

Here n is the electron concentration in CB. For simplicity we are assuming 'A' to be same for both electron – hole recombination centre and also for electron trap centre.

I. At Steady State

Here $\qquad \qquad \dfrac{dn}{dt} = 0$ and $n = n_0$

From equation (1)

$\Rightarrow \qquad \qquad L - An_0\,(n_0 + N) = 0 \qquad \qquad ...(2)$

(A) Low Level (Low Intensity Level)

Here light intensity is low, hence no. of photogenerated carriers is much smaller than the net numbers of trap levels, *i.e.*, $n_0 \ll N$.

So, applying this condition into equation (2),

we have, $\qquad \qquad L - An_0\,(N) = 0$

$\Rightarrow \qquad \qquad n_0 = \dfrac{L}{AN}$

Hence, $\qquad \qquad n_0 \propto L$

(B) High Level (High Intensity Level)

Here light intensity is high, so, $n_0 \gg N$.

So, applying this condition into equation (2) we have,

$$L - An_0\,(n_0) = 0$$

$\Rightarrow \qquad \qquad n_0 = \sqrt{\dfrac{L}{A}}$

Hence, $\qquad \qquad n_0 \propto L^{0.5}$

This is the same expression as obtained in Pure sample.

II. When Light is Just Switched off ($L = 0$)

From equation (1)

$\Rightarrow \qquad \qquad \dfrac{dn}{dt} = 0 - An\,(n + N)$

$\Rightarrow \qquad \qquad \dfrac{dn}{n\,(n + N)} = - A\,dt$

$\Rightarrow \qquad \dfrac{1}{N}\left[\dfrac{1}{n} - \dfrac{1}{n + N}\right] dn = - A\,dt$

Integrating from at $t = 0, n = n_0$ and at $t = t, n = n$, then,

$$\frac{1}{N} \int_{n_0}^{n} \left[\frac{1}{n} - \frac{1}{n+N} \right] dn = -A \int_0^t dt$$

$\Rightarrow$ $\quad\quad\quad$ $[\log (n + N) - \log n]_{n_0}^{n} = ANt$

$\Rightarrow$ $\quad\quad\quad$ $\log \left[\dfrac{n+N}{n} \right] - \log \left[\dfrac{n_0 + N}{n_0} \right] = ANt$ $\quad\quad$ ...(3)

When light is switched off, the light intensity is dying out, hence, we are obviously at low level condition and so $N \gg n_0$ and $N \gg n$. Hence from equation (3),

$\Rightarrow$ $\quad\quad\quad$ $\log \left[\dfrac{N}{n} \right] - \log \left[\dfrac{N}{n_0} \right] = ANt$

$\Rightarrow$ $\quad\quad\quad$ $\log \left[\dfrac{N}{n} \times \dfrac{n_0}{N} \right] = Ant$

$\Rightarrow$ $\quad\quad\quad$ $\dfrac{n_0}{n} = e^{ANt}$

$\Rightarrow$ $\quad\quad\quad$ $n = n_0 e^{-ANt}$ $\quad\quad$ ...(4)

Here the decay of photo carriers having traps follows equation (3).

Response time, t_0 : Now this is defined as the time into which the carriers reduce to $(1/e)$ times of its initial value.

Hence, at $t = t_0, n = n_0/e$. So from equation (3) we have

$\Rightarrow$ $\quad\quad\quad$ $\dfrac{n_0}{e} = n_0 e^{-NAt_0}$

$\Rightarrow$ $\quad\quad\quad$ $e^{-1} = e^{-NAt_0}$

$\Rightarrow$ $\quad\quad\quad$ $NAt_0 = 1$

$\Rightarrow$ $\quad\quad\quad$ $t_0 = \dfrac{1}{NA}$

It shows that in presence of traps, t_0 as well as, conductivity decreases. So more the traps less will be the conduction in the sample. This gives a far more correct picture about t_0 obtained experimentally.

4. PHOTOCONDUCTIVE CELLS

(Materials are Selenium (Se), Lead Sulphide (PbS), Cadmium Sulphide (CdS), etc.)

When light is incident on this material the electrical conductivity of the above materials increases. This increase depends on:

- incident light intensity, and
- frequency of the incident light.

A photoconductive circuit is shown in Figure 4 below.

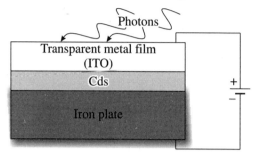

Fig. 4   A cadmium sulophide photoconductor under bias and illumination.

When photon strikes on the surface of the material with $E > E_g$ sufficient energy will be imparted to electron to raise it to CB. Consequently a hole is left in the VB. The EHP is free to serve as current carriers and hence conductivity increases.

Construction

In the photoconductive cell (Fig. 4 above) a thin film of Cadmium Sulphide (CdS) is deposited on one side of an iron or Stainless steel plate. The top surface of the film is coated with a thin film of conducting and transparent oxide, Indium Tin Oxide (ITO). Under the exposure of light current starts flowing in the battery circuit connected between the iron plate and top conducting oxide.

Characteristics

When the cell is not illuminated it shows a resistance in the range ≈ 100 kΩ, this is called Dark Resistance. When illuminated with strong light the cell resistance falls to only a few hundred ohms. Hence the ratio of 'dark' to 'light' resistance of the cell ≈ 1000 : 1. The spectral response of the cell is shown in Fig. 5. Like human eye it is sensitive to visible light. Its spectrum tapers off towards the violet and infrared.

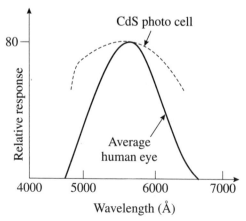

Fig. 5   Spectral response comparison of average human eye and CdS photo cell.

Applications :

1. Light meters,   2. On-Off switch,
3. Street lighting control,   4. Camera exposure setting,
5. Counting   6. Relay control,
7. Aircraft and missile tracking system,
8. Burglar alarm,   9. Voltage regulator.

Advantages :

1. High sensitivity,   2. Low cost,
3. Long life,   4. High dissipation,
5. High voltage (100-300 V)   6. High dark to light resistance.

Disadvantages :

1. Current change with intensity of light with a time lag (high response time)
2. Relatively narrow spectral response.

5. PHOTODIODE

Photodiode is a photosensitive electronic device which produces variable current on incidence of light of variable intensity. When light is allowed to fall on a reverse biased diode *p-n* junction diode additional electron hole pairs are produced due to the absorption of photons. Since majority charge carriers are in large amount as compare to additional charge carriers produced by light incidence. Only minority charge carriers current appreciable changes. These additional minority charge carriers enhance the reverse current. Initially as biasing is increased the current increases linearly and becomes almost constant after certain value. The diode which works on this principle is called as photodiode.

Construction

A photodiode is a *pn* junction of dimension of a few mm embedded in a transparent plastic module. This is shown in Figure below. A lens is placed on the front side of the junction occasionally to enhance the intensity of incident radiation.

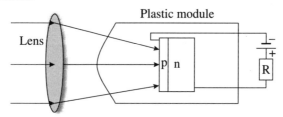

Fig. 6 A schematic diagram of a photo diode under bias under illumination.

Excluding the front side of the capsule all the other sides are enclosed in a metallic casing. The diode is connected in reverse biased with a load resistor *R*.

Principle

Under dark condition and under the application of a large reverse voltage a reverse saturation current is flown through a photodiode. This current (also known as dark current, I_d) is due to the thermally generated minority carriers. This is proportional to the concentrations of minority carriers. Due to a high reverse bias no majority carriers are allowed to flow across the junction. On exposure to light more and more electron hole pairs are formed. The proportional increase of majority carriers under light is much smaller than that of minority carriers. Hence the incident radiation acts as minority carrier injector and by crossing the junction these injected minority carriers contribute to an additional current.

Hence under high reverse bias when diode is subjected to the light net reverse current is

$$I = I_s + I_d$$

Here, I_s is the short circuit current under illumination. Under dark with reverse bias V, the dark current is given by

$$I_d = I_0 [1 - e^{eV/nkT}]$$

Here, I_0 is reverse saturation current, n is the diode ideality factor (for Si, $n = 2$ and for Ge, $n = 1$). Hence the expression for V-I characteristic is given by

$$I = I_s + I_0 [1 - e^{eV/nkT}]$$

Figure represents the I-V curves of a Ge photodiode under different intensity of illuminations.

The dark curve is passing through the origin and the current level increases with the increase in intensity level of illumination. Thus the reverse current can be changed with intensity and it thus acts as light detector.

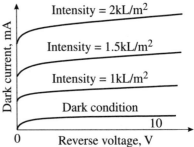

Fig. 7 *I-V* characteristics of a photodiode under different intensity of illumination.

Uses :

1. As a light switch to switch on or off with incident light.
2. Light detectors, light operated switches.
3. Optical communication systems.
4. In all modern instrumentation, control.

6. PHOTOVOLTAIC CELL

Steep rise in the demand of electrical energy in the world makes it necessary to look for alternate non-conventional energy resources other than the rapidly depleted fossil fuels liberating green house gases. Photovoltaic cells or solar cells at present are a sole contender as a reliable renewable energy source. They are the source for electrical power for satellites, space vehicles and for terrestrial applications. Their uses and demand in the world is steadily growing as it generates electricity from light with a minimum or no pollution.

Becquerel in 1839 discovered this phenomenon of development of a voltage or potential difference under the effect of light (photons). This phenomenon is called Photovoltaic effect. Devices based on this effect are known as **Photovoltaic Cells**. A Solar Cell is a device based on Photovoltaic effect.

Construction : The solar cells generally consists of a *pn* junction, where onto its base *p*-region an *n*-region is diffused as shown in Figure 6. It can be other way also. For crystalline silicon solar cells, the bottom side is totally covered and the front side is partially covered by a metal layers. Metal layers are used to collect all the photo generated carriers generated

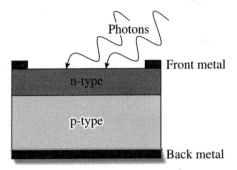

Fig. 8 A photovoltaic cell under illumination (schematic diagram).

into the diode materials. Front side is partially covered ($\approx$ 5 – 7%) to allow more light to incident on the front surface to generate more carriers. The diode materials generally used are semiconductor materials like Gallium Arsenide GaAs), Cadmium Telluride (CdTe), etc. However, most commonly used materials are Silicon, both in single or multi crystalline form.

Principle :

• Solar cell is a device which converts sunlight directly to electricity.

When a solar cell is exposed to sunlight the photons with energy, $h\nu$, greater than E_g of the semiconductor material are absorbed in the cell. In this process a fraction of E_g of photon energy is utilized in creating EHP and the excess energy, $h\nu$-E_g, is dissipated generally in the form of thermal energy given to the crystal.

• Therefore light radiation of λ greater than a certain value of λ is not useful for a solar cell. Such value of λ is known as Cut Off wavelength, λ_g.

In a *pn* junction solar cell the incident photons generate EHP in both *p* and *n* regions of the junction as shown in Figure 9. The EHP thus produced in the vicinity of the junction and in the space charge region (x_s) at the junction is separated by the strong built-in electric field that exists at the junction. This causes the photo generated electrons of the *p*-side to flow due to diffusion; they reach the junction and crossover to the *n*-side. Similarly the photo generated holes of the *n*-side cross over to the *p*-side. This accumulation of electrons on *n*-side and holes on *p*-side of the junction gives rise to a photo voltage.

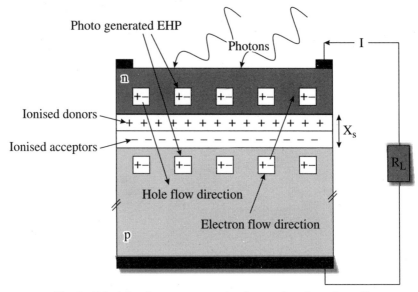

Fig. 9   Principle of energy conversion in a *pn* junction solar cell.

- The photo voltage attains a maximum value when there is infinitely large load across the cell such that the output current of the cell is zero. This is the maximum photo voltage and is known as the **Open Circuit Voltage (V_{oc})**. If a finite load resistance, R_L, is connected across the cell, current flows through it and there is a voltage drop also across the load.
- When the load connected across the diode terminals is zero, the current is maximum and is known as **Short Circuit Current, I_{sc}**.

For a finite value of R_L the current continues to flow in the circuit as long as the solar cell is exposed to sunlight, its magnitude being higher for a higher intensity of light.

Contribution to series resistance, R_s, comes from
- the bulk resistivity of the base material of the cell,
- the bulk resistance of the metallic contacts and interconnections, and
- the contact resistances between the metallic contacts and the semiconductor.

Contribution to shunt resistance, R_{sh} comes from
- the leakage current across the *pn* junction at the peripheral regions and non-peripheral regions due to the presence of (i) defects, (ii) precipitates of foreign impurities, and (iii) non-uniform diffusion. The ideal values of R_s and R_{sh} are zero and infinite respectively. However, in all practical solar cells it is difficult to achieve these values.

Under open circuit condition V_{oc} can be expressed as

$$V_{oc} = \frac{n\kappa T}{q} \ln\left[\frac{I_{sc}}{I_0} + 1\right] \qquad \ldots(1)$$

Here I_0 is reverse saturation current, n is diode ideality factor of the cell, q is the electronic charge, k is Boltzmann's constant and T is the absolute temperature of the cell respectively.

The condition of maximum power is

$$\frac{d(IV)}{dV} = 0 \qquad \ldots(2)$$

The value of maximum power output, P_m, under ideal condition (*i.e.*, $R_s = 0$ and $R_{sh} = \alpha$) is given by

$$P_m = I_m V_m \qquad \ldots(3)$$

Here, I_m and V_m are the values of current and voltage of the cell at maximum power point. The curve factor (CF) or Fill Factor (FF) of the illuminated *I-V* characteristics of the cell is defined as

$$FF = \frac{P_m}{I_{sc} V_{oc}}$$

FF is a measure of how squarish is the *I-V* characteristics of the cell. However, FF is always less than 1. Larger the value of FF more squarish is the *I-V* characteristics of the cell. Increase in R_s and decrease in R_{sh} results in degradation of FF. FF basically denotes the maximum power rectangle inside the curve as also clear from the Figure 10 below.

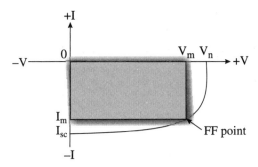

Fig. 10   Actual illuminated *I-V* characteristics of a solar cell.

In terms of V_{oc}, J_{sc} and FF values, the solar cell efficiency, η, can be defined as

$$\eta = \frac{V_{oc} \times J_{sc} \times FF}{P_i} \times 100\%$$

Here, P_i is the input light intensity (power) of the solar cell and J_{sc} is the Short Circuit Current density. Figure 10 above shows an ideal solar cell characteristic curve in fourth quadrant. This is because here current is due to the flow of the minority carriers only and hence is negative. However, for simplicity we draw it on the first quadrant as shown in Figure 11.

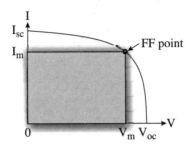

Fig. 11   Converted illuminated *I-V* characteristics of a solar cell.

The value of J_{sc} of a practical solar cell is less than the ideal value. The losses in it are mainly of three types.

- The high refractive index of silicon makes the bare silicon surface highly refractive; it causes ≈ 34% of the incident light to get reflected.
- The metal front contact which generally covers ≈ 5-7% top surface area blocks the same percentage of incoming radiations.
- Some useful light with low absorption coefficient couples out from the cell without being absorbed if the cell is not thick enough.

Applications :

1. Solar Lighting System (Solar Lantern., Solar Home lighting, Solar Street lighting),
2. Sources of power in satellites on space to operate all electronic gadgets.
3. Sources of power in communication system, railway signaling in remote areas.
4. Sensor for optical signals for machines,
5. Solar water pumping,
6. It is the only source of zero ripple D.C. voltage and so best for battery charging.

6

MAGNETIC MATERIALS

Most of the electrical appliances, which are integral part of our daily life such as motors, generators, transformers, loudspeakers, switching circuits, magnetic amplifier, televisions, cassette recorder, telephone diaphragms, food mixer, magnetic core computer memories, magnetic tapes etc., contains iron or its alloy for increasing the magnetic flux without increasing the current. In the last two decades some more new and important applications based on the magnetic properties of materials has come into prominence.

Our understanding of the microstructural factors that influence the magnetic properties is now better than before. The control of microstructure for obtaining the desired magnetic properties is today almost as important as the control necessary for achieving their optimum mechanical properties. The magnetic materials make considerable change in magnetic field by magnetization. Magnetization is the process of alignment of magnetic dipoles with respect of external magnetic field. According to Weber and Ewing each atom of magnetic substances is a tiny bar magnet, having north and south pole. The nature and value of contribution to the magnetization depends on electronic structure of the atom in terms of magnetic moment. This chapter starts with fundamentals associated with magnetization factor. It also discusses the cause of magnetization in magnetic materials. The magnet materials have been classified into various categories depending on the behaviour of material with respect of applied magnetic field. The magnetization factor for various magnetic materials has been described both in qualitative and quantitative way. This helps us to understand the controlling parameter responsible for the magnetization within the material.

1. MAGNETIC FIELD

The magnetic force of attraction is exhibited by two end faces of a bar magnet (north and south pole of the magnet). The magnetic poles always occur in pairs. The space around a magnet where its magnetic influence is experienced is known as Magnetic Field (B).

Fig. 1(a) : It indicates magnetic lines of force around a bar magnet.

Fig. 1(b) : It indicates magnetic dipole moment, μ_m, and it is directed from south to north.

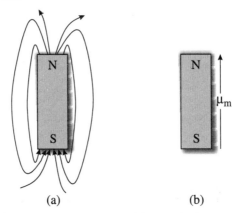

(a)   (b)

Fig. 1   (a) Magnetic lines of forces of a bar magnet
(b) Magnetic dipole moment of a bar magnet.

When current starts flowing through a loop its lines of force and the dipole moment is shown in figure given below.

Fig. 1(c) : It indicates magnetic lines of force around a current loop.

Fig. 1(d) : It indicates magnetic dipole moment associated with current loop.

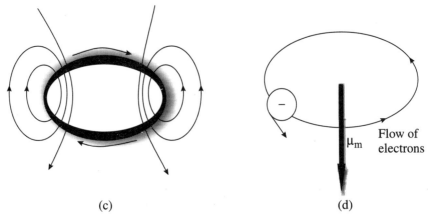

(c)   (d)

Fig. 1   (c) Magnetic lines of forces of a current carrying loop
(d) Magnetic dipole of moment of a current carrying loop.

- Direction of μ_m is determined by the right hand screw rule which states that if the screw is rotated in the direction of current flowing in the loop, the translational motion of the screw will indicate the direction of μ_m.

Some definitions :
- The magnetic lines of force are also known as lines of magnetic induction.
- A line of induction is a curve whose tangent at each point indicates the direction of orientation of a small campus needle.
- The lines of magnetic induction are collectively called Flux.
- The concentration of magnetic flux is called the magnetic flux density and it is a measure of magnetic field strength. It is defined as the number of flux lines passing through a unit area of cross-section.
- The magnetic induction at any point in a magnetic field is the force (F) per unit north pole placed at that point. So, $B = F/m$, where, m is magnetic pole strength.

  Also, B is the force experienced by a moving electric charge in the magnetic field, i.e., $B = F/eV$. Unit of B in SI : Tesla (T) or Weber per square meter (Wb/mt^2) in CGS : Gauss (G), $1\ T = 10,000$ G.

2. | ORIGIN OF MAGNETIZATION

There are three types of magnetic moment associated with each atoms present in solid. They are explained in the following discussions.

(a) Orbital Motion of Electrons

The electronic orbit behaves as an elementary magnet having a magnetic moment. Net orbital magnetic moment is the sum of orbital magnetic moment of individual electrons. The orbital magnetic moment of an electron arises in principle due to their motion around the nucleus. A revolving electron in an orbit can be considered to generate a small current around the nucleus and thus produces a magnetic field.

So its dipole moment is $\mu_{el} = IA$, where, I is the current due to the motion of the electron and A is the area of the orbit. Now if the electron moves in a circular orbit of radius 'r' with a velocity 'v' then its frequency is given by

$$f = \frac{1}{T} = \frac{1}{2\pi r/v} = \frac{v}{2\pi r} = \frac{1}{2\pi}\omega$$

Here T and ω are the time of each revolution and the angular frequency of the electron respectively.

Also $$\omega = \frac{v}{r}.$$

So, $$I = -ef$$

$\Rightarrow$ $$I = -\frac{e}{T} = -e \cdot \frac{\omega}{2\pi}$$

and $$A = \pi r^2$$

As $\qquad \mu_{el} = IA$

$\Rightarrow \qquad \mu_{el} = \left(-\dfrac{e}{2\pi} \times \omega\right) \times \pi r^2$

$\Rightarrow \qquad \mu_{el} = -\dfrac{e\omega r^2}{2} = -\dfrac{1}{2}evr$

$\Rightarrow \qquad \mu_{el} = \left(-\dfrac{e}{2m}\right)[mvr]$

$\Rightarrow \qquad \mu_{el} = \left(-\dfrac{e}{2m}\right)L$

Here, L is the orbital angular momentum. From Bohr's postulate of quantized orbit, angular momentum of electron is

$$L = n \times \dfrac{h}{2\pi}, \qquad \text{where, } n = 1, 2, 3, \dots$$

Hence, $\qquad mvr = \dfrac{nh}{2\pi}$

$\Rightarrow \qquad vr = \dfrac{nh}{2\pi m}$

So, $\qquad \mu_{el} = -\dfrac{1}{2}evr = -\dfrac{1}{2}e \times \dfrac{nh}{2\pi m}$

$$\mu_{el} = -n\left(\dfrac{eh}{4\pi m}\right)$$

Thus the magnetic moment of electron due to its orbital motion is an integral multiple of $\dfrac{eh}{4\pi m}$. This quantity '$\dfrac{eh}{4\pi m}$' is a natural unit for magnetic moment of electron and is called the '**Bohr Magneton**', μ_B.

Bohr Magneton,

$$\mu_B = \dfrac{eh}{4\pi m} = \dfrac{(1.9 \times 10^{-19}) \times (6.63 \times 10^{-34})}{4 \times 3.14 \times (9.1 \times 10^{-31})} = 9.28 \times 10^{-24} \text{ A-mt}^2$$

In quantum mechanics, $L = \dfrac{h}{2\pi}\sqrt{l\,(l+1)}$, where, $l = 0, 1, 2, \dots$ (for s, p, d, orbitals)

(b) Electron Spin

When electron revolves around the nucleus in different orbits it also spins around its own axis. This time electron behaves like tiny bar magnet. The spin magnetic moment associated with this spinning motion is given by $\mu_s = -g\mu_B s$. Here μ_B is the Bohr Magneton, s is the spin quantum number and $s = 1/2$. The electronic intrinsic spin angular momentum, S, is given by

$$S = \dfrac{h}{2\pi}\sqrt{s\,(s+1)}.$$

(c) Nuclear Spin

In an atom the nucleus of atom also rotates about the centre of mass of the atom, i.e., electron and nucleus rotate about centre of mass of the atom. Since the nucleus is heavy, in comparison to electron, its magnetic moment is very small. Its magnetic moment is $\approx \dfrac{1}{2000}$ of an electron, and thus can be neglected. So when such atoms are placed in a magnetic field the orbital motion of electrons undergoes changes and the atoms acquire an induced magnetic moment.

3. MAGNETIC DIPOLE MOMENT (μ_m)

Case-I : Bar Magnet

A bar magnet is freely suspended in a uniform magnetic field of B, it experiences a torque which tends to align the magnet with the field. So the force acting on each pole is mB (from the definition of B). The condition of the magnet is depicted in the Fig. 2(a) given below. The equal and opposite forces at north and south poles constitutes a couple. The moment of this couple is $mB \times d \sin \theta$, where d is the distance of separation between poles and θ is the angle that the axis of the magnet makes with B. Hence, torque, τ, is

$$mB \times d \sin \theta = md \cdot B \sin \theta = \mu_m B \sin \theta$$

So, $$\vec{\tau} = \vec{\mu}_m \times \vec{B} \qquad \ldots(1)$$

Here, μ_m is the magnetic dipole moment of the bar magnet (pole strength multiplied by distance between the poles).

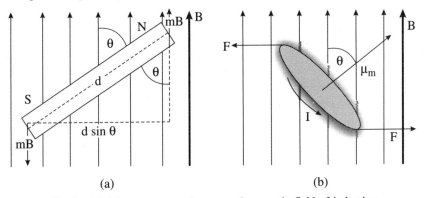

Fig. 2 (a) A bar magnet under external magnetic field of induction
(b) A current carrying loop under external magnetic field of induction.

Case-II : Current Carrying Loop

When a loop is freely suspended in the same way of bar magnet it also gets deflected due to the force acting on it as shown in figure below. The torque, τ, acting on a loop carrying a current I is directly proportional to the area of the loop A. Also, $\tau \propto I$, and $\tau \propto B$.

Hence, $$\tau = IAB \sin \theta = \mu_m B \sin \theta$$

$$\Rightarrow \qquad \vec{\tau} = \vec{\mu}_m \times \vec{B} \qquad \ldots(2)$$

Here, μ_m is the magnetic dipole moment of the loop and is equal to 'IA'.

Magnetization (M) :

When a solid is placed in a magnetic field the magnetic lines of force are redistributed. The solid somehow enhances the magnetic induction B.

$$\therefore \qquad B = B_0 + B_{solid}$$

where, B_0 is the primary field, B_{solid} is the field created by the solid.

$$\therefore \qquad B = \mu_0 H + \mu_0 M \qquad \qquad ...(3)$$

Here $B_{solid} = \mu_0 M$

This is the extra magnetic induction due to the solid and M is magnetization of solid. Each element in the volume of the solid behaves as a small magnet and the magnetic moment of the solid is the vector sum of the magnetic moment of such element. M can be defined as the induced dipole moment per unit volume, i.e.,

$$M = \frac{\sum\limits_{\Delta V_i} \mu_{m_i}}{V}$$

where, ΔV_i is the elementary physical volume and V is the total volume of the solid.

Magnetic Susceptibility (X_m) :

Now, $M \propto H \implies M = X_m H$ ...(4)

Here, X_m is the proportionality constant and is known as Magnetic Susceptibility of the material medium.

- X_m is defined as the ease with which the material can be magnetized. X_m is positive for Paramagnetic and Ferromagnetic material, and negative for Diamagnetic material

Magnetic Permeability (μ) :

From equation (3), $B = \mu_0 (H + M)$

$\implies$ $B = \mu_0 (H + X_m H)$

$\implies$ $B = (1 + X_m) \mu_0 H$

$\implies$ $B = \mu_0 \mu_r H$ ...(5)

Here, $\mu_r = 1 + X_m$ ...(6)

Here, μ_r is the relative permeability of the material and $\mu_r > 1$ or $\mu_r < 1$.

Now, as $B = \mu H$, so from equation (5)

$\implies$ $\mu = \mu_0 \mu_r$ ...(7)

Here, μ is the permeability of the medium and

$$\mu_r = \frac{\mu}{\mu_0} \qquad \qquad ...(8)$$

- It indicates the extent that magnetic field lines permeate through the medium.

Relationship between μ_r and χ :

When the specimen is placed in a magnetic field H, it is magnetized due to the alignment of the current loops. The magnetic flux density within the specimen is the sum of the magnetizing field and the induced magnetic field. So the intensity of magnetization is given as,

$$B = B_0 + \mu_0 M$$

So,
$$\mu H = \mu_0 H + \mu_0 M$$

$\Rightarrow$
$$\frac{\mu}{\mu_0} = 1 + \frac{M}{H}$$

$\Rightarrow$
$$\mu_r = 1 + X \text{ (as } \mu_r = \mu/\mu_0 \text{ and } X = M/H)$$

$\Rightarrow$
$$X = \mu_r - 1$$

4. MAGNETIC MATERIALS

Magnetic materials are generally classified into five categories. Those are the followings.

   (i) Diamagnetic,   (ii) Paramagnetic,

  (iii) Ferromagnetic,   (iv) Ferrimagnetic, and

  (v) Antiferromagnetic.

However, we are going to discuss here about the first three major magnetic materials.

I. Diamagnetic

Here in absence of any applied field, dipoles do not exist as also shown in Fig. 3(a) given below. Under the application of field the dipoles are induced opposite to that of applied field direction as shown in Fig. 3(b) below. This strength of the induced magnetic field is proportional to the intensity of applied magnetic field.

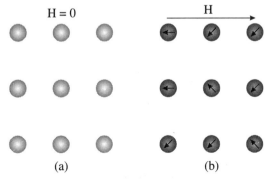

Fig. 3 (a) Diamagnetic atomic dipoles under no external magnetic field
(b) Diamagnetic atomic dipoles under external magnetic field.

Characteristics :

(i)   When placed in a magnetic field it acquires feeble magnetism in a direction opposite to that of the applied field. Here portion of the material near N-pole becomes N-side and portion of the material near S-pole becomes S-side.

(ii)   It shows negative X_m as directions of M is opposite to that of H and so $|X_m| \approx 10^{-6}$.

(iii) As $X_m < 0$, so $\mu_r < 1$.

(iv) If a diamagnetic material is placed in an inhomogeneous magnetic field it tends to be pushed into the regions of weaker field as in Figure 4(a) given below.

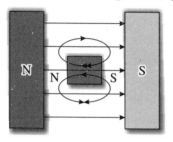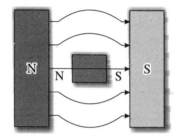

Fig. 4(a)   Induced magnetization in diamagnet under external magnetic field H.

(v) X_m is practically independent of temperature.

(vi)   When the field is not so strong, M is a linear function of H.

(vii) **Examples :** Au, Ag, Bi, Cu, Sb, Air, Water, H_2, CO_2, N_2, etc.

(viii) When diamagnetic liquid is placed into a magnetic field, the liquid depress at the centre as magnetic field is stronger at the centre as shown in Figure 4(b).

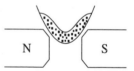

Fig. 4(b) A diamagnetic liquid under external magnetic field.

II. Paramagnetic

Here molecules possesses a net permanent dipole moment even in $H = 0$ as also seen from Figure 5(a) shown below. However, due to their randomness in direction, the net magnetization of the material is zero. Under the application of external field 'H' the dipoles tend to align in the direction of applied field and the material becomes magnetized as shown in Figure 5(b) shown below.

- With the increase in temperature the more thermal agitation tends to randomize the dipole directions leading to a decrease in magnetization.

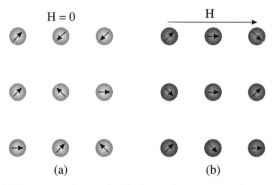

Fig. 5 (a) Paramagnetic atomic dipoles under no external magnetic field
(b) Paramagnetic atomic dipoles under external magnetic field.

Characteristics :

(i) When placed in a magnetic field it acquires feeble magnetism in a direction of the applied magnetic field. Here portion of the material near N-pole becomes S-side and portion of the material near S-pole becomes N-side.

(ii) It shows positive value of X_m as directions of M is same as that of H and so $X_m \approx 10^{-6}$.

(iii) As $X_m > 0$, so $\mu_r > 1$. So field lines are pulled towards the materials and permeate through it when placed in a magnetic field as shown in Figure 6(a) below.

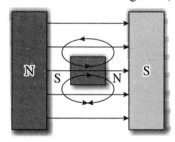

Fig. 6(a) Induced magnetization in paramagnet under external magnetic field.

(iv) When a rod of paramagnetic substance is freely suspended in a magnetic field it aligns itself along the lines of induction. In a non-uniform field the paramagnetic substances are attracted towards stronger region of magnetic field.

(v) Variation of M with H is linear when H is not too strong.

(vi) X_m is strongly dependent on temperature.

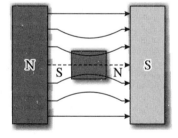

Fig. 6(b) A paramagnetic liquid under external magnetic fielde.

Roughly, $X_m \propto \dfrac{1}{T} \Rightarrow$ This is **Curie Law.**

(vii) **Examples :** O_2, $CuCl_2$, Cr, Pt, solutions of Fe-salts, Al, Mg, Mn, etc.

(viii) When paramagnetic liquid is placed into a magnetic field, the liquid accumulate at the centre as shown in Figure 6(b).

III. Ferromagnetic

The atoms of these materials posses net magnetic moment. Permanent magnetic dipoles are already present even in the absence of any external magnetic field. Under the influence of external magnetic field all the dipoles become strongly magnetized in the direction of applied magnetic field.

- Unlike the dia and paramagnetism experiments they show Spin magnetic moments of electrons responsible for ferromagnetism.
- The ferromagnetic materials exhibit spontaneous magnetization and a spontaneous magnetic field even in the absence of an applied magnetic field.

Characteristics :

(i)   It becomes strongly magnetized when placed in a magnetic field in the direction of the field.

(ii) X_m is very high, $X_m \approx 10^6$.

(iii) $\mu_r \gg 1$, $\mu_r \approx 1000$.

(iv) M does not vary linearly with H. The variation of B with H is shown in Figure 7(a) below.

(v)   Field lines crowd into the material like the paramagnetic one, but with large intensity as shown in Figure 7(b) below.

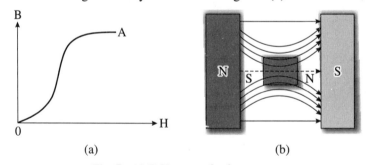

(a)   (b)

Fig. 7   (a) B-H curve of a ferromagnet
(b) Induced magnetization in ferromagnet under external magnetic field.

(vi) Ferromagnetic properties of crystals are direction dependent. **Example :** Fe : easily magnetized along <100>, but not along <111>.

(vii) Above a particular temperature (Curie temperature, T_C) ferromagnetic properties of the material disappears.

Material	T_C (°K)
Fe	1043
Co	1404
Ni	631

(viii) It shows **hystersis** as shown in Figure 7(c) below.

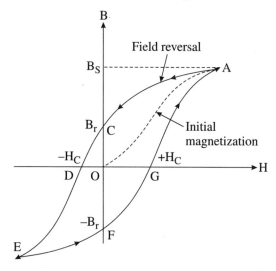

Fig. 7 (c) Hysteresis curve of a ferromagnet.

(ix) Examples are Fe, Co, Ni, some types of steel, etc. An unmagnetized sample is started getting magnetized. It follows the path *OA*. With a reversal of magnetic field it follows path *ACDE* and if again getting magnetized it follows path *EFGA*.

- At zero field, *i.e.*, $H = 0$, $B = B_r$ at point *C*. This is called the Remanance or remanant flux density $\Rightarrow$ **Retentivity**.
- At point *A*, $B = B_s$ $\Rightarrow$ **Saturation flux density**.
- At $B = 0$ (or to make $B = 0$), $H = -H_C$ at *D*. H_C $\Rightarrow$ **Coercivity or the coercive force**, The above $B - H$ curve is known as **the Hystersis Loop**.

5. CLASSICAL THEORY OF DIAMAGNETISM (LANGEVIN)

When a substance is subjected to a magnetic field there is a weak alignment of magnetic dipoles in the opposite direction to the applied magnetic field. This means a development of weak magnetization in the specimen in opposite direction of the applied magnetic field. This phenomenon is known as diamagnetism. Diamagnetism is present in all substances, but, generally suppressed by the larger effects due to permanent magnetic dipole moments of the atoms. Langevin gave the explanation of this phenomenon by a new phenomenon known as 'Larmor precession'. In the absence of any magnetic field, electron of the atom revolves around the nucleus with a certain angular velocity (natural frequency). When a magnetic field is applied to the substance, angular frequency of the electron changes. The associated orbital angular momentum or magnetic moment of electron starts precessing about the field direction. This precession is known as Larmor precession. The angular frequency changes in such a way that the magnetic dipoles which were aligned in the field direction, their magnitude of angular frequency decreases. The other dipoles which were aligned opposite to the field direction, their magnitude of angular frequency increases. So the net effect is a weak alignment of magnetic dipoles in opposite to the applied magnetic field.

Let us consider an electron revolves around the nucleus. If 'm' and '$-e$' are the mass and charge of electron respectively and 'r' is the radius of the orbit, then, in absence of an external magnetic field, the centripetal force acting on the electron due to nucleus is given by $F_0 = \dfrac{mv^2}{r} = m\omega_0^2 r$.

Here 'v' and 'ω_0' are the linear and angular velocities of electron respectively. When the magnetic field is applied along the Z-axis, an additional force, F_L, equal to the Lorentz force starts to act on the electron in direction away from the centre. Due to change in the net force on electron there is a change in the angular frequency from 'ω_0' to 'ω'. Actually in the absence of magnetic field the electron motion being spherically symmetrical produces no net current or flux. On the other hand on application of external magnetic field the electron's motion is no longer spherically symmetric, but precesses about the field and produces a net current 'I'.

Now the Lorentz force on the electron is then given by
$$F_L = -e(v \times B) = -er\omega B$$
The resulting equation of motion is given by
$$mr\omega^2 = m\omega_0^2 r - eBr\omega$$
$$\Rightarrow \qquad \omega^2 = \omega_0^2 - \frac{eB}{m}\omega$$
$$\Rightarrow \qquad \omega^2 + \left(\frac{eB}{m}\right)\omega - \omega_0^2 = 0$$
The roots of above equation are then
$$\omega = \pm\left[\left(\frac{eB}{2m}\right)^2 + \omega_0^2\right]^{1/2} - \frac{eB}{2m}$$
Since $\omega_0 \gg \dfrac{eB}{m}$, then, $\left(\dfrac{eB}{2m}\right)^2$ term can be neglected.

So, $$\omega = \pm\,\omega_0 - \frac{eB}{2m} = \pm\,\omega_0 - \omega_L$$

where $\omega_L = \dfrac{eB}{2m}$ is called the Larmor frequency.

- **Larmor theorem** states that for an atom in a magnetic field the motion of the electron is to the first order in 'B', the same as a possible motion in absence of 'B' except for the superposition theorem of a precession of the electrons with angular frequency, ω_L.

The **precession (Larmor)** of magnetic moment 'μ_{el}' in a magnetic field is equivalent to the precession of angular momentum L about 'B'. The $\pm$ sign on ω_0 indicates that those electrons whose orbital moments are parallel to the field are slowed down by ω_L and those electrons whose orbital moments are anti-parallel to the field are speeded up by ω_L.

So in one case we have $\omega = \omega_0 - \omega_L$ and in the other case we have $\omega = -\omega_0 - \omega_L$. The current, I, is given by

$$I = \text{charge} \times \text{revolutions per sec.} = -\frac{e\omega_L}{2\pi} = \frac{e^2 B}{4\pi m}.$$

For an atom with atomic number, Z, current becomes

$$I = -Z\frac{e^2 B}{4\pi m}$$

So magnetic moment of this loop is then

$$\mu = \text{current} \times \text{area} = -\frac{Ze^2 B}{4\pi m} \times \pi \langle a^2 \rangle \qquad \ldots(1)$$

Here, $\langle a^2 \rangle$ is the average radius of the electron from the field axis.

The negative sign indicates that the induced magnetic moment is always in opposite sense to that of applied field.

If the field is in Z-direction,

$$\langle a^2 \rangle = \langle x^2 \rangle + \langle y^2 \rangle \qquad \ldots(2)$$

But, the mean square distance of the electrons from the nucleus is,

$$\langle r^2 \rangle = \langle x^2 \rangle + \langle y^2 \rangle + \langle z^2 \rangle \qquad \ldots(3)$$

For a spherically symmetric charge distribution,

$$\langle x^2 \rangle = \langle y^2 \rangle = \langle z^2 \rangle$$

Hence,

$$\langle a^2 \rangle = 2 \langle x^2 \rangle \Rightarrow \langle x^2 \rangle = \langle a^2 \rangle / 2$$

So,

$$\langle x^2 \rangle = \langle y^2 \rangle = \langle z^2 \rangle = \frac{\langle a^2 \rangle}{2}$$

$$\Rightarrow \qquad \langle r^2 \rangle = \frac{3 \langle a^2 \rangle}{2} \qquad \text{[using equation (3)]} \quad \ldots(4)$$

From equations (1) and (4),

$$\therefore \qquad \mu = -\frac{Ze^2 B}{4m} \times \frac{2}{3} \times \langle r^2 \rangle$$

where, N is the numbers of atoms per unit volume, then, $M = \mu N$.

$$\therefore \qquad M = -\frac{Ze^2 NB}{6m} \langle r^2 \rangle$$

Susceptibility, $\qquad X_{\text{dia}} = \frac{M}{H} = -\frac{Ze^2 N\mu_0}{6m} \langle r^2 \rangle$

This is the classical Langevin equation for a diamagnetic material. It is clear that X_{dia} is proportional to N implying that its dependence on the size of the atom. Also X_{dia} is independent of temperature. Experimentally, $X_{dia} \approx 10^{-6}$ and it is thereby a small effect.

6. | CLASSICAL THEORY OF PARAMAGNETISM (LANGEVIN)

Langevin (1905) considered that the magnetic moments arising from the orbital motion of electrons causes paramagnetic behaviour and he represented the paramagnetic material as 'a gas of magnetic needles'. Here 'gas' emphasizes that the interaction between magnetic moments is neglected. In absence of an external magnetic field the magnetic axes of molecules are uniformly distributed in all directions to have net magnetization zero on the macroscopic level.

The potential energy, PE, corresponding to a magnetic dipole in a magnetic field of induction B is

$$W = -\overrightarrow{\mu_m} \cdot \overrightarrow{B} = -\mu_m B \cos\theta$$

where 'θ' is the angle between the directions of B and the magnetic dipole.

This potential energy, W, attains its minimum value when the direction of dipole orientation coincides with the direction of B, *i.e.*, $\theta = 0$. Since for a stable system W should be minimum, the magnetic dipole moment tends to orient along external magnetic field direction. So when a paramagnetic material is subjected to a magnetic field, the action of the field turns the dipoles so that they are aligned with the field. Thermal agitation, however, opposes the ordering action of magnetic field. The degree of orientation of magnetic dipoles, thus, is dependent on both the magnetic field strength and the temperature of the material.

Theory

We now consider a hard sphere of unit radius consisting of N dipoles, each having dipole moment μ_m. Thus N will represent numbers of dipoles for unit volume in the material. We also assume that they are in thermal equilibrium at a temperature T. At any instant, the number of magnetic dipoles dN or $N(\theta)\, d\theta$ that lie within a solid angle $d\Omega$, is proportional to the solid angle, which is given by $2\pi \sin\theta\, d\theta$.

This is also shown in Figure 8.

$\therefore$ $\qquad\qquad N(\theta)\, d\theta \propto 2\pi \sin\theta\, d\theta$

$\Rightarrow$ $\qquad\qquad dN \propto 2\pi \sin\theta\, d\theta$ $\qquad\qquad$ ...(1)

Now we suppose that the gas of magnetic moment is subjected to a magnetic field of induction B.

According to Maxwell-Boltzmann (MB) statistics, the number of dipoles with PE is W and is proportional to the Boltzmann's factor $e^{-W/kT}$; k is the Boltzmann's constant.

$\therefore$ $\qquad\qquad N \propto e^{-W/kT}$ $\qquad\qquad$ ...(2)

Hence using eqns. (1) and (2), the number of dipoles dN which have PE as W and lying within the solid angle $d\Omega$ is given by

$$dN = A_0 e^{-W/\kappa T} d\Omega$$

$\Rightarrow$ $$dN = A_0\, e^{-(-\mu_m B \cos \theta)/\kappa T} \cdot (2\pi \sin \theta\, d\theta)$$

$\Rightarrow$ $$dN = 2\pi A_0 \cdot e^{\mu_m B \cos \theta/\kappa T} \cdot \sin \theta\, d\theta$$

Here A_0 is constant of proportionality.

So, $$dN = A \cdot e^{\beta \cos \theta} \cdot \sin \theta\, d\theta \qquad \qquad ...(3)$$

Here, $$A = 2\pi A_0 \quad \text{and} \quad \beta = \frac{\mu_m B}{kT}$$

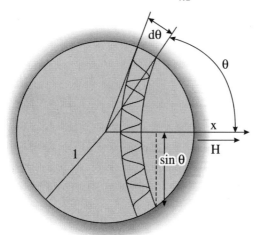

Fig. 8 Schematic diagram of a unit sphere of a paramagnetic material.

So, the total number of dipoles within unit volume having PE 'W' is

$$N = \int_0^\pi dN = \int_0^\pi A \cdot e^{\beta \cos \theta} \sin \theta\, d\theta \qquad ...(4)$$

As each dipole carries a magnetic moment μ_m, the net magnetic moment contributed by dN dipoles is

$$dp = dN \times \mu_m \cos \theta$$

The total magnetization is

$$M = \int_0^\pi dp = \int_0^\pi dN \times \mu_m \cos \theta$$

$\Rightarrow$ $$M = A\mu_m \int_0^\pi e^{\beta \cos \theta} \cos \theta \sin \theta\, d\theta$$

[using equation (3)] ...(5)

Also from equation (4),

$$A = \frac{N}{\displaystyle\int_0^\pi e^{\beta \cos \theta} \sin \theta\, d\theta} \qquad ...(6)$$

Hence from equations, (5) and (6),

$$M = N\mu_m \frac{\int_0^\pi e^{\beta\cos\theta} \cos\theta \sin\theta \, d\theta}{\int_0^\pi e^{\beta\cos\theta} \sin\theta \, d\theta}$$

$$M = N\mu_m \frac{\int_{+1}^{-1} e^{\beta y} \cdot y \cdot (-dy)}{\int_{+1}^{-1} e^{\beta y} \cdot (-dy)}$$

by substituting $\cos\theta = y$, on differentiating,
$\Rightarrow -\sin\theta \, d\theta = dy$. At $\theta = 0$, $y = +1$ and at $\theta = \pi$, $y = -1$.

So, $$M = N\mu_m \frac{\left[\frac{e^{\beta y}}{\beta} \cdot y - \int 1 \cdot \frac{e^{\beta y}}{\beta} \cdot dy\right]_{+1}^{-1}}{\left[\frac{e^{\beta y}}{\beta}\right]_{+1}^{-1}}$$

$\Rightarrow$ $$M = N\mu_m \frac{\left[(-1)\left(\frac{e^{-\beta}}{\beta}\right) - (+1)\left(\frac{e^{+\beta}}{\beta}\right)\right] - \frac{1}{\beta^2}[e^{-\beta} - e^{+\beta}]}{\frac{1}{\beta}[e^{-\beta} - e^{+\beta}]}$$

$\Rightarrow$ $$M = N\mu_m \frac{-\frac{1}{\beta} \cdot 2 \cdot \left[\frac{e^{-\beta} + e^{+\beta}}{2}\right] + \frac{1}{\beta^2} \cdot 2 \cdot \left[\frac{e^{+\beta} - e^{-\beta}}{2}\right]}{-\frac{2}{\beta}\left[\frac{e^{+\beta} - e^{-\beta}}{2}\right]}$$

$\Rightarrow$ $$M = N\mu_m \frac{\cosh\beta - \left(\frac{1}{\beta}\right)\sinh\beta}{\sinh\beta}$$

Hence, $$M = N\mu_m \left[\coth\beta - \left(\frac{1}{\beta}\right)\right] = N\mu_m \cdot L(\beta) \qquad \ldots(7)$$

where, $$L(\beta) = \left[\coth\beta - \left(\frac{1}{\beta}\right)\right] \Rightarrow \textbf{Langevin's Function} \qquad \ldots(8)$$

Case-I : When $\mu_m B \gg \kappa T$ and $\beta \gg 1$:

$\therefore$ $$|L(\beta)|_{\beta\to\infty} = \left|\coth\beta - \frac{1}{\beta}\right|_{\beta\to\infty}$$

$\Rightarrow$ $$|L(\beta)|_{\beta\to\infty} = \left|\frac{e^{\beta} + e^{-\beta}}{e^{\beta} - e^{-\beta}} - \frac{1}{\beta}\right|_{\beta\to\infty}$$

$\Rightarrow$ $$|L(\beta)|_{\beta\to\infty} = \left|\frac{e^{\beta} + 0}{e^{\beta} - 0} - 0\right| = 1$$

$$\therefore \qquad\qquad M = N\mu_m \Rightarrow M_s$$

This is the saturation value of Magnetization.

This will be attained when B is very large and T is low.

Case-II : When $\mu_m B \ll \kappa T$ and $\beta \ll 1$:

Value of μ_m of an atom $\approx 10^{-23}$ J/T.

At moderate field, i.e., $B = 1$ T, $\mu_m B \approx 10^{-23}$ J,

But, $kT \approx 4 \times 10^{-21}$ J at room temperature.

$\therefore \qquad\qquad \mu_m B \ll \kappa T$ for moderate B.

Now for $\beta \ll 1$, $\qquad L(\beta) = \coth \beta - \dfrac{1}{\beta}$

$$\Rightarrow \qquad\qquad L(\beta) = \left[\frac{1}{\beta} + \frac{\beta}{3} - \frac{\beta^3}{45} + \frac{2\beta^5}{945} + ...\right] - \frac{1}{\beta}$$

$$\Rightarrow \qquad\qquad L(\beta) \approx \left[\frac{1}{\beta} + \frac{\beta}{3}\right] - \frac{1}{\beta}$$

$$\Rightarrow \qquad L(\beta) = \frac{\beta}{3} = \frac{\mu_m B}{3\kappa T}$$

$$\therefore \qquad\qquad M = N\mu_m \left[\frac{\mu_m B}{3\kappa T}\right]$$

$$\Rightarrow \qquad\qquad M = \frac{N\mu_m^2}{3\kappa}\left[\frac{B}{T}\right]$$

$$\Rightarrow \qquad\qquad M = \frac{N\mu_m^2 \mu_0 H}{3\kappa T} \qquad ...(9)$$

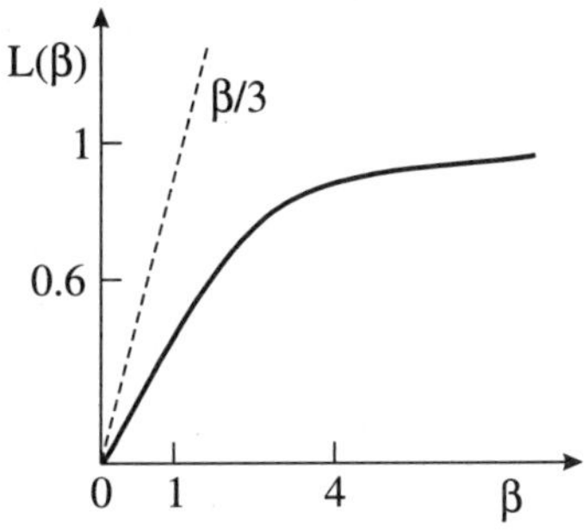

Fig. 9 The plot of $L(\beta)$ with β for $\beta \ll 1$.

The variation of $L(\beta)$ with β is shown in Figure 9 given aside.

Hence, paramagnetic susceptibility, X_{para}, is

$$X_{para} = \frac{M}{H} = \frac{N\mu_m^2 \mu_0}{3\kappa T} = \frac{C}{T} \qquad ...(10)$$

Here, **Curie constant**, C, is given by

$$C = \frac{N\mu_m^2 \mu_0}{3\kappa} \qquad ...(11)$$

Also, from eqn. (9)

$$M \propto \frac{B}{T}$$

It indicates that increase in 'B' tends to align magnetic moment so that 'M' increases, whereas, 'T' tends to decrease the alignment.

• For the attainment of saturation value 'M_s' we should have high magnetic field as 50 KG (5 T) and a temperature of 1.3 K.

The variation of '$1/\chi$' with T and the variation of 'χ' with T are shown in Figure 10(a) and 10(b) respectively.

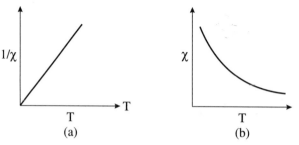

Fig. 10   (a) The $1/\chi$ vs T graph for a paramagnet
(b) The χ vs T graph for a paramagnet.

Some paramagnetic substances do not follow this relation, as in this theory it is considered that there is no interaction between neighbours which build their own internal molecular field. Weiss modified this theory after consideration of above mentioned fact.

7. WEISS THEORY OF FERROMAGNETISM

(A) Spontaneous Magnetization

Weiss postulated that in ferromagnetic materials the internal molecular field is responsible for lining up of dipoles in the same direction.

The internal field is given by

$$H_i = H + \gamma M \qquad \qquad ...(1)$$

Here, H is applied field and 'γM' is a measure for the tendency of the environment to align a given dipole parallel to the magnetization already existing. 'γ' is known as the Molecular Constant or the Weiss constant. The equation (1) represents the Curie-Weiss Law and the occurrence of spontaneous magnetization. It is assumed that ferromagnetics are essentially paramagnetic materials having a very large molecular field. So in case of the paramagnetic state of a ferromagnetic material,

$$X_{para} = \frac{M}{H_i} \text{ and also } X_{para} = \frac{C}{T}$$

$$\Rightarrow \qquad \frac{M}{H + \gamma M} = \frac{C}{T}$$

$$\Rightarrow \qquad MT = CH + \gamma MC \Rightarrow M = \left[\frac{CH}{T - \gamma C}\right]$$

$$\Rightarrow \qquad M = \left[\frac{CH}{T - \gamma C}\right] \qquad \qquad ...(2)$$

Hence, $$\qquad X_{para} = \frac{C}{T - \gamma C} = \frac{C}{T - \theta} \qquad \qquad ...(3)$$

Equation (3) is identical to Curie-Weiss law, where, $\theta = \gamma C$.

• When $\theta \rightarrow T$, then from eqn. (3), $X_{para} \rightarrow \infty$.

It indicates that the interactions of the individual magnetic moments reinforce each other causing them to align parallel at $T = \theta$.

Physical origin of Weiss Molecular Field

The introduction of Weiss Molecular Field made it possible to explain a wide range of phenomenon observed in ferromagnetic. Initially ferromagnetic property was considered to be due to nearest neighbour dipole – dipole interaction. Such an interaction would give a field interaction of order 10^3 gauss only. While experimentally it is found to be 10^7 gauss. Thus a much stronger interaction type is needed to explain this phenomenon.

An attempt was made to explain this phenomenon by Heisenberg in 1928 to explain large Weiss Molecular Field by so called exchange interaction between the electron spins. The exchange energy for ions i, j bearing spins S_i, S_j can be expressed as

$$E_{ex}^{ij} = -2 J \vec{S_i} \cdot \vec{S_j}$$

and the total exchange energy of spin i with its nearest neighbours $J_1, J_2 \ldots$ may be expressed as

$$E_{ex}^{ij} = -2 J \sum_{j=1}^{n} \vec{S_i} \cdot \vec{S_j}$$

Here J is called the exchange integral and is a measure of strength of interaction.

In the above equations, the dot product is used because exchange energy is governed by relative orientation of spins. In ferromagnetic, where parallel orientation of spins are favoured E_{ex} is minimum and $\vec{S_i} \cdot \vec{S_j} = S^2$.

E_{ex} is maximum when spins are antiparallel and therefore

$$\vec{S_i} \cdot \vec{S_j} = -S^2$$

The exchange integral can be plotted as a function of the interatomic spacing 'a' to the unfilled shell of radius 'r'.

According to Slater (1930), the ratio $(a/r) \geq 3$ (but not very large) must be satisfied for the occurrence of ferromagnetism. The substances viz. Fe, Co, Ni and Gd satisfy the above criteria and are found to be ferromagnetic. On the other hand, Cr and Mn fail to satisfy the above criterion and are not ferromagnetic. However they can also be made ferromagnetic by allowing them with some suitable nonferromagnetic elements (with slightly higher but comparable interatomic spacing) to form compounds whose ratio (a/r) is greater than 3. For example, Mn-As, Cu-Mn, Mn-Sb show ferromagnetic behaviour because of favourable (a/r) ratio.

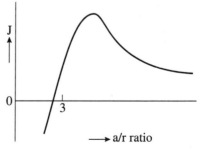

Fig. 11 The variation of exchange integral, J with a/r ratio for a ferromagnet.

Discrepancy of γ Values

Experimentally, $\gamma \approx 10^3$, however, it is 1000 times larger than the theoretical value.

Reason : We considered interaction between dipoles as classical magnetic interactions, whereas, actually the interaction is of quantum mechanical one and the forces between dipoles are 'Exchange Forces'.

Equation (1) shows that for $T = \theta$, $M = \dfrac{H_i}{\gamma}$, when $H = 0$.

That is why the magnetization at $T \leq \theta$ is spontaneous.

Expression for Spontaneous Magnetization

When $$H = 0, H_i = \gamma M$$

$\therefore$
$$\beta = \frac{\mu_m H_i}{\kappa T} = \frac{\mu_m \gamma M}{\kappa T}$$

$\Rightarrow$
$$M = \beta \left[\frac{\kappa T}{\gamma \mu_m} \right] \qquad \ldots(4)$$

Also from earlier,
$$M = N\mu_m \left[\coth \beta - \left(\frac{1}{\beta} \right) \right] \qquad \ldots(5)$$

Both equations (4) and (5) give the condition for spontaneous magnetization. Figure 12 given below shows the plot between M and β. Straight lines show the linear relationship as per eqn. (4) at different T. The other curve shows the Langevin curve as per eqn. (5). The intersection point A of a given temperature line with the Langevin curve represents the finite spontaneous magnetization at that temperature. If the dipoles assume state C, then the local magnetization is less than the corresponding equilibrium state A. The magnetization and Langevin variable β will increase until the state A is reached. With an increase in temperature the straight lines increase in slope as in eqn. (4) which brings

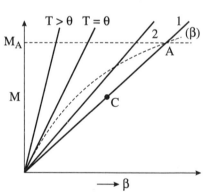

Fig. 12 Langevin function $L(\beta)$ and M vs β curves at four different temperatures.

down the point of intercept A downwards as in curve 2. Finally at $T = \theta$, *i.e.*, at Curie temperature, there is no intercept and there is no spontaneous magnetization.

At $T = \theta$ the slope of the straight line is identical to the slope of the Langevin curve near the origin.

Hence, $\dfrac{M}{\beta}$ [from eqn.(4)] $= \dfrac{M}{\beta}$ [from eqn.(5)]

So,
$$\frac{k\theta}{\gamma\mu_m} = \frac{N\mu_m}{3} = \frac{M_s}{3}$$

[when $\mu_m\beta \ll \kappa T$ and $\beta \ll 1$, from equation (5)

$\Rightarrow$ $\qquad\qquad L(\beta) = \frac{\beta}{3}, M = \frac{N\mu_m\beta}{3}, M_s = N\mu_m$]

$\Rightarrow$ $\qquad\qquad\qquad \gamma M_s = \frac{3\kappa\theta}{\mu_m}$

This is known as Weiss Molecular Field or Weiss Field.

The condition for stable spontaneous magnetization is $T < \theta$. So below the Curie point 'θ' the material is spontaneously magnetized depending upon the temperature. Above 'θ' the ferromagnetic material turns to a paramagnetic one.

(B) Domain Hypothesis

This hypothesis of Weiss does explain why a ferromagnetic material has no net dipole moment.

Postulates

The entire ferromagnetic volume is splitted into a large number of small regions of spontaneous magnetization. These regions are called the 'Domains'. Within each domain the magnetic moments are oriented parallel to one another and every domain is characterized by a definite value and direction of magnetic field. In absence of any external field the magnetic moment vectors are oriented randomly as in Figure 13(a) earlier so that the net magnetic moments of the entire volume is zero. Domains are the microscopical region in the substance in which all the atomic dipoles are aligned in one direction. When the magnetic field is applied, the domains rotate to make them align their magnetic moments with the field directions as in Figure 13(b) instead of the individual atomic dipoles attempting to line up parallel to the field.

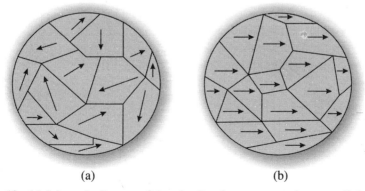

(a)   (b)

Fig. 13   (a) Schematic diagram of domains in a ferromagnet under no applied field
(b) Schematic diagram of domains under applied field.

Explanation of Hystersis using Domains

In absence of a magnetic field the domains in the material are randomly oriented and the net magnetic moment is zero. When the material is placed in a magnetic field H, the orientations of the magnetization vectors of various domains with respect to the magnetic field direction are different as shown in Figure 14. The magnetization vector of the 1st domain forms the smallest

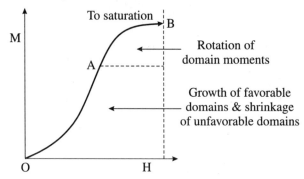

Fig. 14   Magnetization graph under applied field showing domain growth and domain rotation.

angle with H. With an increase of H, the growth of most favorably oriented domain '1' grows at the expense of its neighbour domains by sidewise motion of domain walls. Domain '1' grows until the whole crystal becomes a single domain.

- With an increase of H, M proceeds along OAB. OA corresponds to the wall motion. A further increase in H beyond point A causes the rotation of domain moments into the field direction. M reaches at this stage the technical saturation.

- Beyond point 'B' the growth of magnetization with H is very slow. In the single domain formed under the action of the magnetic field the individual magnetic moment alignment would not be complete due to the disturbing effect of temperature. An increasing H prevails over the thermal disturbance and tends to orient the atomic dipoles. A single crystal is considered to be made up of a large number of domains. On application of applied magnetic field after the specimen acquires saturation magnetization, all the magnetic domains then aligned parallel to the direction of applied field.

7

SUPERCONDUCTIVITY

The fascinating term of zero resistivity of superconductors make this the most wanted materials of the last century. Within a certain critical limits these materials practically show zero resistivity along with its perfect diamagnetism. First property makes them a potential candidate for carrying electrical power of undiminished strength over very large distances, making very high power magnets etc. The anti-gravitation or levitation property of a diamagnetic material like superconductors becomes important for other use. Till now large amount of money had been spent in its research to have room temperature superconductor and research is still on. Here we are going to discuss some important basic theories of superconductors and their uses.

1. INTRODUCTION

Superconductivity is one of the most fascinating physical phenomena having a vast potentiality in practical applications. In 1908 Dutch physicist Kammerlingh Onnes succeeded in his efforts to liquify Helium gas at 4.2 K at atmospheric pressure. Using liquid helium as coolant the variation of electrical resistance of metals at low temperature was studied by Onnes. In 1911 Onnes discovered that the electrical resistivity of highly purified mercury dropped abruptly to nearly zero at 4.15 K as shown in Fig. 1 aside. This sudden drop of resistance was not in accordance with the expectations and was a completely new phenomenon 'Superconductivity'. Later on this phenomenon was observed in Pb, Sn, Zn, Al and other metals.

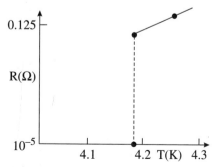

Fig. 1 Variation of resistance of mercury with absolute temperature.

- The sudden disappearance of electrical resistance in materials below a certain temperature is known as '**Superconductivity**'.
- The materials which exhibit superconductivity and which are in 'Superconducting' state are called '**Superconductors**'.
- The temperature at which a normal materials turn into a superconductor is called the '**Critical temperature, T_C**'.
- Every superconductors have its own 'T_C' at which it passes over into the superconducting state. A sudden fall in resistance for semiconductors indicates the transition to the superconducting state as shown in Figure 2. Till 1973 the highest T_C was known for Nb_3Ge

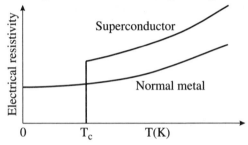

Fig. 2   Variation of electrical resistivity with absolute temperature for normal metal and superconductors.

alloy as 23.2 K. In 1986 a T_C of 35 K was reported for La-Ba-Cu-O system. Till date many researchers have found $T_C > 100$ K in many superconducting systems. The big saving in the cost of these applications is due to the fact that liquid nitrogen (boiling point : 77 K) could be used to cool them rather than liquid helium (boiling point : 4.2 K), then the era of High T_C superconductor commenced. Good electrical conductors are not superconductors. On the other hand the superconductors are very poor conductors at room temperatures. Berilium (Be), aluminium (Al), niobium (Nb), molybdenum (MO) and zinc (Zn) exhibit superconductivity. Semiconductors like Si, Ge, selenium (Se) and tellurium (Te) become metal at very high pressure and then at low temperature become superconductors.

- Those metals which are very good conductors at room temperatures, *i.e.*, Ag, Au, Cu, etc., are not at all superconductors even at 0.05 K.
- Some semiconductors (or chemical compounds) became superconductors at low temperatures.
- No general criteria for determining a material to become superconductors or not, has been established. The superconducting properties can change by varying (i) temperature, (ii) magnetic field, (iii) stress, (iv) impurity, (v) atomic structure, (vi) size, (vii) frequency of excitation of applied electric field, and (viii) isotope mass.

2. ▌ EFFECT OF MAGNETIC FIELD

If a magnetic field is applied parallel to the length of the superconducting wire, the resistance of the wire is suddenly restored at a finite value of the magnetic field strength and hence the superconductivity will disappear. This magnetic field is known as 'Critical Field' and depends on temperature. This is denoted by $H_C(T)$. This restoration of resistance is abrupt if the metal is perfectly pure and free from stains. Thereby at $T = T_C$, $H_C(T_C) = 0$. The variations of H_C with T are shown in Fig. 3 for different materials. The curves are nearly parabolic and can be represented by

$$H_C = H_0 \left[1 - \frac{T^2}{T_C^2} \right],$$

where, H_0 is the critical field at $T = 0$ K.

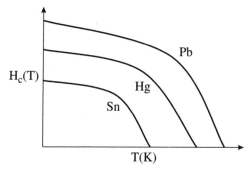

Fig. 3   The variation of critical magnetic field with absolute temperature.

1. Persistent current : When we start a current to pass through a close loop of superconducting material, the current will continue to keep flowing on its own till the temperature of the loop kept at below the critical temperature. This current of constant magnitude is known as persistent current. It has been found theoretically that once the current flow is started it will continue for more than 10^5 years. This is very useful for superconducting magnets as it does not require any power supply to maintain its magnetic field.

2. Heat Capacity : For any superconducting metal its transition below critical temperature to the superconducting state does not involve any crystallographic structural changes. This is only a thermodynamic phase change. Here te specific heat (C_v) changes discontinuously at the critical temperature. This variation of C_v is given by $C_v = e^{-bT_c/T}$, where, b is a constant.

3. Isotope Effect : Reynolds and Maxwell observed the variation of T_C with isotopic mass, M, as

$$T_C \propto M^{-1/2} \Rightarrow T_C M^{1/2} = \text{constant.}$$

3. FLUX EXCLUSION : THE MEISSNER EFFECT

In an ordinary conductor the electric field is given by $E = \dfrac{1}{\sigma} J = \rho J$, where J is the current density and ρ is resistivity of the conductor. So for a perfect conductor, $\rho = 0$. Hence, $E = 0$.

If such a perfect conductor is placed in a magnetic field and the magnetic field is withdrawn, the changing magnetic flux will not induce an e.m.f (e) in it.

From Faraday's Law,

$$e = -\frac{d\phi}{dt}.$$

So when $e = 0$, ϕ becomes constant. Thus the magnetic flux through the perfect conductor remains unchanged and the applied magnetic field is frozen or trapped in the perfect conductor. Also if a perfect conductor is taken into a magnetic field the magnetic field lines are pushed aside as shown in Fig. 4. A superconductor behaves in same way as with perfect conductor when placed into a magnetic field. It does not allow field lines to penetrate into it.

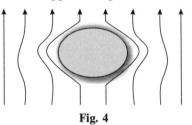

Fig. 4

However, in 1933 Meissner found that a superconductor completely expels any magnetic field lines that were initially penetrating in the normal state. The magnetic flux was totally expelled from the sample when cooled beyond T_C.

- This expulsion of magnetic flux during the transition from the normal to the superconducting state is called the **Meissner Effect**.

It shows that for superconductors both $\dfrac{dB}{dt} = 0$ and $B = 0$, independent of the path by which the state is reached.

Perfect Diamagnetism

A superconductor behaves as a perfect diamagnetic material as $B = 0$.

So, $\qquad\qquad B = \mu_0 (H + M) = 0$

$\Rightarrow \qquad\qquad M = -H$

Hence, magnetic susceptibility,

$$X = \frac{M}{H} = -1$$

Because of diamagnetic nature all superconducting materials strongly repel external magnets. This gives rise to both the **Levitation Effect,** and **Suspension Effect.**

Here a piece of superconductor hangs beneath a magnet. This is a sure test for superconductor.

4. LONDON EQUATIONS

To explain Meissner effect in superconductors London (and London) in 1935 proposed a modification of the electromagnetic wave equations. The new equations besides explaining Meissner effect also explain observations on thin film superconductors. These equations are known as London equations. It was assumed that there are two type of electrons, namely 'normal' and 'super electron' in superconductors. Let n_n and n_s are the densities of normal and super electrons respectively, the conduction electron 'n' becomes

$$n = n_n + n_s$$

The normal current and super current are assumed to flow in parallel. Super current flows without resistance, while normal electron flow remain quite inert and are thus ignored in the superconducting state.

The force equation for super electrons is

$$m \frac{d\vec{v}}{dt} = -e\vec{E} \qquad \text{...(1)}$$

Here, m and v are the mass and velocity of electron respectively, $-e$ is electronic charge and E is the applied electric field.

Now, current density in the sample is

$$\vec{J} = -ne\vec{v} \qquad \text{...(2)}$$

Here, n is no. of electrons per unit volume and here $n = n_s$.

Hence, on differentiation with respect to time from equation (2),

$$\frac{d\vec{J}}{dt} = -ne \frac{d\vec{v}}{dt}$$

$$\Rightarrow \qquad \frac{d\vec{J}}{dt} = -ne\left(-\frac{e\vec{E}}{m}\right)$$

$$\Rightarrow \qquad \frac{d\vec{J}}{dt} = \frac{ne^2}{m}\vec{E} \qquad \text{...(3)}$$

For $E = 0$, J becomes constant. This is the basic relation describing the absence of resistance and is known as the 1st London equation. It states that no electric field is necessary unless the current changes, steady currents may be set up in the absence of electric field. This is the phenomenon of 'Superconductivity'. Superconductors consist of both normal and superconducting electrons, we are considering only the superconducting electrons here.

From Maxwell's equations,

$$\vec{\nabla} \times \vec{E} = -\frac{d\vec{B}}{dt} \qquad \text{...(4)}$$

and

$$\vec{\nabla} \times \vec{H} = +\vec{J} \qquad \text{...(5)}$$

Taking curl of both sides of eqn. (3),

$$\vec{\nabla} \times \frac{d\vec{J}}{dt} = \frac{ne^2}{m}(\vec{\nabla} \times \vec{E}) = \frac{ne^2}{m}\left(-\frac{d\vec{B}}{dt}\right) \qquad \text{(using equation (4))}$$

$$\Rightarrow \quad \vec{\nabla} \times \left(A\frac{d\vec{J}}{dt}\right) = \left(-\frac{d\vec{B}}{dt}\right), \qquad \qquad [\text{where, } A = \frac{m}{ne^2}] \quad \dots(6)$$

On differentiation of eqn. (5) with respect to t,

$$\frac{d\vec{J}}{dt} = \vec{\nabla} \times \frac{d\vec{H}}{dt}; \qquad \qquad \dots(7)$$

Substituting (7) into (6),

$$\vec{\nabla} \times \left(A \cdot \vec{\nabla} \times \frac{d\vec{H}}{dt}\right) = -\mu_0 \frac{d\vec{H}}{dt} \qquad (\text{as } \vec{B} = \mu_0\vec{H})$$

$$\Rightarrow \qquad A \cdot \vec{\nabla} \times \left(\vec{\nabla} \times \frac{d\vec{H}}{dt}\right) = -\mu_0 \frac{d\vec{H}}{dt}$$

$$\Rightarrow \qquad A\vec{\nabla} \cdot \left(\vec{\nabla} \cdot \frac{d\vec{H}}{dt}\right) - A\nabla^2\left(\frac{d\vec{H}}{dt}\right) = -\mu_0 \frac{d\vec{H}}{dt}$$

$$\Rightarrow \qquad A\vec{\nabla} \cdot \left(\frac{d(\vec{\nabla} \cdot \vec{H})}{dt}\right) - A\nabla^2\left(\frac{d\vec{H}}{dt}\right) = -\mu_0 \frac{d\vec{H}}{dt}$$

$$\Rightarrow \qquad 0 - A\nabla^2\left(\frac{d\vec{H}}{dt}\right) = -\mu_0 \frac{d\vec{H}}{dt} \qquad (\text{as } \vec{\nabla} \cdot \vec{H} = 0)$$

$$\Rightarrow \qquad A\nabla^2\left(\frac{d\vec{H}}{dt}\right) = \mu_0 \frac{d\vec{H}}{dt} \qquad \dots(8)$$

Integrating with respect to time,

$$\nabla^2 A\,(\vec{H} - \vec{H_0}) = \mu_0\,(\vec{H} - \vec{H_0}) \qquad \dots(9)$$

Here, H_0 is the constant of integration and also the field at $t = 0$. Equation (9) is a direct consequence of Maxwell's equation. However, from Meissner Effect, at T_C we have zero magnetic field, *i.e.*, at $t = 0$, $H_0 = 0$. So according to Londons we should eliminate $\vec{H_0}$.

Hence instead of equation (6) the correct equation will be by omitting time derivative, as,

$$\vec{\nabla} \times (A \cdot \vec{J}) = -\vec{B} = -\mu_0\vec{H} \qquad \dots(10)$$

This is the fundamental equation in a superconductor as it gets rid of $\vec{H_0}$ term. This is also called the (second) **London equation**. Sometimes equation (10) along with equation (3) together are known as **London equations**.

5. EXPLANATION OF MEISSNER EFFECT AND FLUX PENETRATION FROM LONDON EQUATIONS

From Maxwell's as eqn. (5),

$$\vec{\nabla} \times \vec{H} = \vec{J} \Rightarrow \vec{\nabla} \times \vec{B} = \mu_0 \vec{J}$$

Taking curl of both sides,

$$\vec{\nabla} \times (\vec{\nabla} \times \vec{B}) = \mu_0 (\vec{\nabla} \times \vec{J})$$

$$\Rightarrow \qquad \vec{\nabla} . (\vec{\nabla} . \vec{B}) - \nabla^2 \vec{B} = \mu_0 (\vec{\nabla} \times \vec{J})$$

$$\Rightarrow \qquad \nabla^2 \vec{B} = \mu_0 \left[-\frac{ne^2}{m} \vec{B} \right]$$

(from $\vec{\nabla} . \vec{B} = 0$ and Eqn. (10))

$$\Rightarrow \qquad \nabla^2 \vec{B} = \frac{ne^2 \mu_0}{m} \vec{B}$$

$$\Rightarrow \qquad \nabla^2 \vec{B} = \frac{1}{\lambda^2} \vec{B} \qquad \qquad ...(11)$$

where, $\lambda = \dfrac{1}{\sqrt{\dfrac{ne^2\mu_0}{m}}}$

Equation (11) is a simple differential equation with a solution

$$B = B_0 \, e^{-x/\lambda} \qquad \text{in one dimension.} \quad ...(12)$$

Here, B_0 is magnetic flux density at surface and x is the distance from the surface into the specimen. The solution in equation (12) agrees quite well with the experimental data. Experimental decrease of the magnetic field inside a superconductor is also shown in Fig. 5 aside.

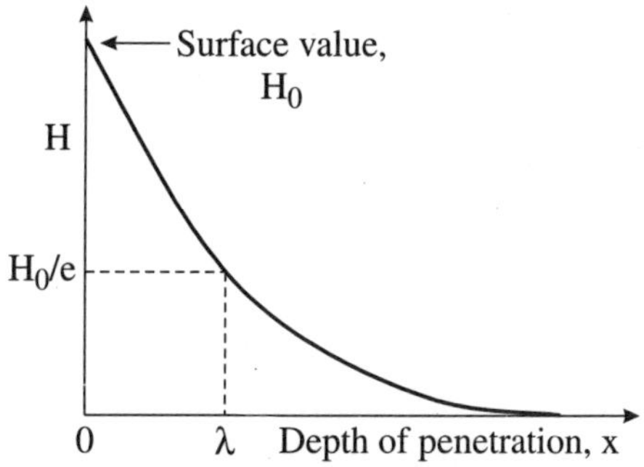

Fig. 5 Penetration of magnetic flux lines on application of external magnetic field.

- Above equation shows that the magnetic field does not drop abruptly to zero into the specimen, but decays inside the superconductor exponentially from the surface through a distance (characteristic of the material) λ, called **London Penetration depth**.

- The flux density decays exponentially in a superconductor falling $(1/e)$th of its value at the surface at a distance λ, called **London Penetration depth**.

6. | JOSEPHSON EFFECT

The electronic wave function has always a finite probability for quantum mechanical tunneling through a barrier. Normally, significant tunneling is possible for those electrons whose de-Broglie wavelengths are comparable to or greater than the barrier thickness (normally, 50 Å or less). In 1962 Josephson showed that tunneling of superconducting electrons (*i.e.*, Cooper pairs) was possible in the same way of the tunneling of normal electrons. Esaki, Giaver and Josephson received the Nobel prize in 1973 for their work on electron tunneling.

Figure 6(a) showed a schematic arrangement of Josephson junction. It consists of two superconductors separated by a thin (10–20 Å) barrier insulating layer. As shown in Figure 6(b) when a dc current flows through

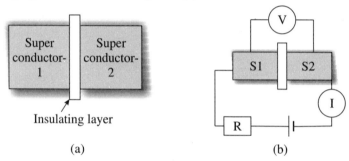

(a)   (b)

Fig. 6 (a) Schematic diagram of a Josephson junction
(b) Explanation of Josephson effect in a Josephson junction.

this junction, the potential difference across its ends is zero. The current flows both ways, one through the superconductor and other across the junction. This is known as the D.C. Joshephson Effect. The current flows by tunneling of Cooper pairs from one superconducting film into the other while penetrating the barrier.

The super current passing through the junction is given by

$$I_S = I_C \sin \theta$$

Here, θ is the phase difference between the wave functions which describe the Cooper pairs on the both sides of the barrier interface, I_C is the critical current supported by the junction and is dependent on thickness, width of the interface. If I_S exceeds I_C large amount of current passes through the junction and yields a potential difference, V, between the two superconducting films (normally then the barrier potential between the Cooper pairs of both sides is ≈ 2 eV). This causes a frequency difference, ν and is given by

$$\nu = \frac{2\,eV}{h} \qquad\qquad ...(1)$$

and, phase difference is given by,

$$\theta = 2\pi\nu t = 2\pi t \left(\frac{2\,eV}{h} \right) \qquad\qquad ...(2)$$

Hence using the equations (1) & (2) we have,

$$I_S = I_C \sin\left(\frac{4\pi \, eV}{h} t\right) \qquad \ldots(3)$$

This current is an alternating current and it exists only when a dc voltage exists across the junction. This is known as the AC Josephson Effect.

When $V = 1 \, \mu V$, an ac current of frequency 483.6 MHz is developed across the junction. Due to ac Josephson effect we have now a new definition of volt.

One volt is now defined as the potential difference when applied across a Josephson junction to generate an electromagnetic radiation of 483597.9 GHz.

7. CLASSIFICATIONS OF SUPERCONDUCTORS (BASED ON THEIR MAGNETIC FIELD BEHAVIOR)

(a) Type-I

They obey complete Meissner Effect, *i.e.*, they are perfect diamagnetic as shown in Figures below. The variation of resistivity and magnetization with externally applied magnetic field are shown in these Figures. Pure specimen of many materials shows this behavior. Below H_C the specimen is superconducting abruptly and the magnetic field fully absent. They are often called soft because of their tendency to give away to low magnetic fields.

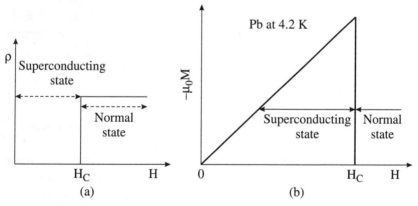

Fig. 7   (a) Variation of resistivity with applied magnetic field
(b) Variation of magnetisation with applied magnetic field.

(b) Type-II

These materials exhibit a magnetization as shown in Figures 8(a) & 8(b) below and are also known as hard superconductors. The variation of resistivity and magnetization with externally applied magnetic field are shown.

For $H < H_{C1}$, the material is diamagnetic and $B = 0$. H_{C1} is known as the Lower Critical Field. At H_{C1} the flux begins to penetrate the specimen

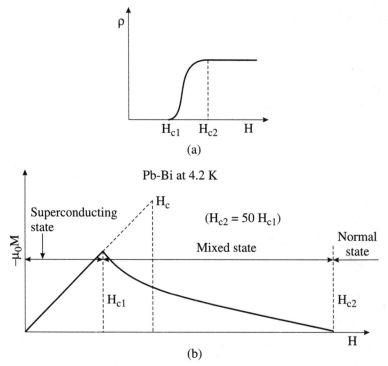

Fig. 8 (a) Variation of resistivity with applied magnetic field
(b) Variation of magnetization with applied magnetic field.

and penetration increases until H_{C2} is reached. Here $M = 0$ and specimen becomes a normal material. This H_{C2} is known as the Upper Critical Field. Between H_{C1} and H_{C2}, $B \neq 0$ and the Meissner effect is said to be complete. There is a gradual fall in $(- M)$ as H increases in Type-II. In this region between H_{C1} and H_{C2} the superconductivity is partially destroyed and is known as 'Mixed State'.

8. ELEMENTS OF BCS THEORY

The modern theory of superconductivity was put forward by Bardeen, Cooper and Schrieffer in 1957 and hence name as 'BCS theory'. This theory successfully explained all observable effects such as zero resistivity, the Meissner effect, isotope effect and some other phenomenon. Important aspects of this theory is discussed below in brief and qualitative manner.

(i) **Electron–Electron Interaction Via Lattice Deformation :** In order to understand the direct interaction, let us consider an electron passing through the packing of positive ions. It is attracted by the neighbouring positive ions and gets screened by them. This screening reduces effective charge of electron. At the same time due to attraction between the electron and ion core, the lattice gets deformed on local scale. This deformation is greater for smaller mass of positive ions core. Second electron passing through this assembly see deformed lattice and gets attracted towards the assembly. This interaction is called a electron-electron interaction via lattice deformation. In the language of field theory, the above interaction is said to be due to exchange of virtual phonon "q" between two electrons. In terms of the wave vector k of the two electrons the process can be written as $k_1 - q = k_1'$ and $k_2 + q = k_2'$ which gives net wave vector $(k_1 + k_2 = k_1 + k_2')$ as zero.

(ii) Cooper Pair : At absolute zero temperature all quantum states with energies $E \leq E_f$ are filled by electrons. In this situation, Cooper showed that if there is an attraction between the two electrons they are able to form a bound state so that their total energy is less than $2E_f$. These electron are paired to form a single system and their motion are correlated. These two electrons together form a Cooper pair and is known as a Cooper electron. These pair of electrons is in fact super electrons which are responsible for superconductivity.

(iii) BCS Ground state : In superconducting materials excitation is different from normal metals. In this case when a pair of electron lying just below the Fermi surface is taken just above it, they form a Cooper pair and their total energy is reduced, it can be done for other pairs until the system can gain no additional energy by pair formation and hence the total energy is further reduced.

9. APPLICATIONS

(i)   Generation and transmission of electricity (saves lots of power loss),

(ii)   Medical diagnostics, *e.g.*, MRI, NMR,

(iii)   Enormously powerful electromagnets,

(iv)   Supercomputers and information processing,

(v)   Magnetically levitating the world's fastest trains,

(vi)   Magnetic energy storage,

(vii)   Electromagnetic shielding,

(viii) Superconducting transformers.

NUMERICALS

Q.1. Find the Miller indices of a set of parallel planes which makes intercepts as $3a : 4b$ on the X and y-axes, and, are parallel to Z-axis. Here $\vec{a}, \vec{b}, \vec{c}$ are the primitive vectors of the lattice.

[**Ans.** (430)]

Hints : $h : k : l = \dfrac{a}{p} : \dfrac{b}{q} : \dfrac{c}{r} = \dfrac{1}{3} : \dfrac{1}{4} : \dfrac{1}{\infty} = 4 : 3 : 0$

Q.2. In an ortho-rhombic crystal a lattice plane cuts intercepts of lengths $3a, -2b, 3c/2$ along the three axes where $\vec{a}, \vec{b}, \vec{c}$ are the primitive vectors of the lattice. Deduce the Miller Indices of the plane.

[**Ans.** $(2, -3, 4)$].

Hints : $h : k : l = \dfrac{a}{p} : \dfrac{b}{q} : \dfrac{c}{r} = \dfrac{a}{3a} : -\dfrac{b}{2b} : \dfrac{2a}{3c}$

Q.3. In a crystal whose primitives are 1.2 Å, 1.8 Å and 2 Å, a plane whose Miller Indices are (231) cuts intercept 1.2 Å along X-axis. What will be the lengths of intercepts along Y and Z-axis?

[**Ans.** $l_2 = 1.2$ Å, $l_3 = 4$ Å].

Hints : $2 : 3 : 1 = \dfrac{1.2}{p} : \dfrac{1.8}{q} : \dfrac{2}{r}$

If l_1, l_2, l_3 are the actual intercepts on X, Y and Z axes respectively,

then $l_1 : l_2 : l_3 = 3.6 : 3.6 : 12 \Rightarrow l_1 = 1.2$ Å $= 3.6\lambda \Rightarrow \lambda = \dfrac{1}{3}$ Å;

$l_2 = 3.6\lambda = 1.2$ Å, $l_3 = 12\lambda = 4$ Å.

Q.4. In a simple cubic lattice find the ratio of intercepts on the three axes by $(1\bar{3}2)$ planes.

[**Ans.** $l_1 : l_2 : l_3 = 6a : -2a : 3a = 6 : -2 : 3$].

Hints : $l_1 : l_2 : l_3 = pa : qb : rc$ In simple cubic lattice $a = b = c$;

& $\qquad \dfrac{1}{p} : \dfrac{1}{q} : \dfrac{1}{r} = h : k : l$

Q.5. **Lead is an FCC lattice with an atomic radius of 1.746 Å. Find the spacing of (i) (200) planes, (ii) (220) planes.**

[**Ans.** $d_{200} = 2.465$ Å, $d_{220} = 1.748$ Å]

Hints : $d_{h, k, l} = \dfrac{a}{\sqrt{h^2 + k^2 + l^2}}$, As lead is F.C.C., $a = \dfrac{4r}{\sqrt{2}} = \dfrac{4 \times 1.746}{\sqrt{2}}$

(i) $d_{200} = \dfrac{4.93}{\sqrt{2^2 + 0 + 0}}$ Å, $d_{220} = \dfrac{4.93}{\sqrt{2^2 + 2^2 + 0}}$ Å

Q.6. **In a general lattice $a = b = 2.5$ Å, $c = 1.8$ Å. Deduce lattice spacing between (111) planes.**

[**Ans.** 1.26 Å]

Hints : $d_{h, k, l} = \dfrac{1}{\sqrt{\dfrac{h^2}{a^2} + \dfrac{k^2}{b^2} + \dfrac{l^2}{c^2}}}$

Q.7. **In a SCC (i) find the ratio of intercepts on the three axes by (123) plane, (ii) Find the ratio of the spacing of the (110) and (111) planes, (iii) Find the ratio of the nearest neighbour distance to the next nearest neighbour distance.**

[**Ans.** $(6 : 3 : 2)$, $d_{110} : d_{111} = \sqrt{3} : \sqrt{2}$, $(1 : \sqrt{2})$]

Hints : (i) In simple cubic lattice

$\dfrac{1}{p} : \dfrac{1}{q} : \dfrac{1}{r} = h : k : l \Rightarrow p : q : r = \dfrac{1}{h} : \dfrac{1}{k} : \dfrac{1}{l} \Rightarrow p : q : r = \dfrac{1}{1} : \dfrac{1}{2} : \dfrac{1}{3}$

(ii) $d_{h, k, l} = \dfrac{a}{\sqrt{h^2 + k^2 + l^2}}$; $d_{110} : d_{111} = \sqrt{3} : \sqrt{2}$

(iii) distance of nearest neighbour is $d_1 = a$, distance of next nearest neighbour is $d_2 = a\sqrt{2}$. So $d_1 : d_2 = 1 : \sqrt{2}$.

Q.8. **Sodium crystallizes as a cubic lattice. The edge of the unit cell is 4.3 Å. The density of Sodium is 963 Kg/mt^3 and its atomic weight is 23. How many atoms are contained in a unit cell ? What type of cubic unit cell does sodium form?**

[**Ans.** 2 atoms, BCC]

Hint : $\rho = \dfrac{MN_{eff}}{a^3 N_A}$; where N_{eff} is the effective number of atoms per unit cell.

Q.9. **Sodium Chloride crystallizes in FCC structure with density 2160 Kg/mt^3. If the atomic weight of sodium is 23 and of chlorine is 35.5, calculate the distance between the two adjacent atoms.**

[**Ans.** 2.815 Å]

Hints : $a = 4.3$ Å, $\rho = 963$ Kg/m^3, $M_{Na} = 23$ Kg/Kmol,

$M_{Cl} = 35.5$ Kg/mol, $N_A = 6.023 \times 10^{28}$ atoms/kmol

$$\rho = \frac{(M_{Na} + M_{Cl}) \times N_{eff}}{a^3 \times N_A}$$

Q.10. The spacing between successive (100) planes in NaCl is 2.82 Å. X-ray incident upon the surface of this crystal and it is found that it gives rise to the first order Bragg reflection at a grazing angle of 8°35′. Calculate the wavelength of X-ray and find the angle at which the 2nd order Bragg reflection would occur.

[**Ans.** 0.842 Å, 17°22′].

Hints : $2d \sin \theta = n\lambda$

Q.11. The Bragg angle corresponding to the 1st order reflection from plane (111) in a crystal is 30° when X-rays of $\lambda = 1.75$ Å are used. Calculate the interatomic spacing.

[**Ans.** 3.03 Å]

Hints : $2d \sin \theta = n\lambda$

$$d_{h, k, l} = \frac{a}{\sqrt{h^2 + k^2 + l^2}}$$

Q.12. Calculate the glancing angle on the cube (110) of rock salt crystal ($a = 2.814$ Å) corresponding to 2nd order diffraction maximum for X-ray with $\lambda = 0.71$ Å.

[**Ans.** 20°55′].

Q.13. Electrons are accelerated to 344 volts and are reflected from a crystal. The 1st reflection maximum occurs when glancing angle is 60°. Determine the spacing of the crystal. Given, $h = 6.62 \times 10^{-34}$ J-sec, $e = 1.6 \times 10^{-19}$ Coulomb, $m = 9.1 \times 10^{-31}$ Kg.

[**Ans.** 0.38 Å]

Hints : $\lambda = \dfrac{h}{\sqrt{2mE}} \Rightarrow \lambda = \dfrac{h}{\sqrt{2m\,eV}}$

$$2d \sin \theta = n\lambda$$

Q.14. If *PE* is expressed as $U = -\dfrac{A}{R^6} + \dfrac{B}{R^{12}}$, show that (i) the intermolecular distance R_0 is given by $\left[\dfrac{2B}{A}\right]^{1/6}$ and

(ii) $U_{min} = -\dfrac{A^2}{4B}$.

Hints : $\dfrac{dU}{dR}\Big|_{R=R_0} = 0$, $U_{min} = U|_{R=R_0}$

Q.15. The mutual interaction potential of two particle system is expressed as $U = -\dfrac{A}{R^m} + \dfrac{B}{R^n}$. Show that the *PE* of these two particles in stable configuration is equal to $\dfrac{4}{5}\left[\dfrac{A}{R_0^2}\right]$ for $m = 2$, $n = 10$.

Q.16. The fraction of vacancy sites in a metal is 1×10^{-10} at 500°C. What will be the fraction of vacancy sites at double the temperature?

[**Ans.** 8.45×10^{-7}]

Hints : $f = \dfrac{n}{N} = e^{-E_p/2kT}$

Q.17. The density of Schottkey defects in a certain sample of NaCl is 5×10^{11} per mt^3 at 25°C. If the observed interionic ($Na^+ - Cl^-$) distance is 2.18 Å, what is the average energy required to create one Schottkey defect?

[**Ans.** 1.971 eV].

Hints : $V = (2a)^3$, $N_{eff} = 4$, $n = \dfrac{N_{eff}}{V}$, $n = Ne^{-E_s/2kT}$

Q.18. The average energy required to create Frenkel defect in an ionic crystal is 1.4 eV. Calculate the ratio of the number of Frenkel defects at 20°C & 300°C for 1 gram of the crystal.

[**Ans.** 1.33×10^{-6}]

Hints : $n = (NN_i)^{1/2} e^{-E_F/2kT}$

Q.19. For a SCC lattice compare the density of lattice points in (111) & (110) planes.

[**Ans.** $\sqrt{2} : \sqrt{3}$].

Hints : $\rho = \dfrac{d_{hkl}}{a^3}$

Q.20. Polonium belongs to SCC lattice. If the lattice constant is 3.36 Å, calculate its density. (The atomic mass of Polonium is 209).

[**Ans.** 9150 kg/mt^3]

Hints : $\rho = \dfrac{N_{eff}\, M}{N_A\, a^3}$

Q.21. The potential energy function for the force between two atoms in a diatomic molecule is $V(x) = -\dfrac{b}{x^6} + \dfrac{a}{x^{12}}$, where a & b are positive constants, x is the distance between the two atoms. (i) At what value of x, $V(x)$ is equal to zero? (ii) At what value of x, $V(x)$ is minimum? (iii) Derive an expression for the force between two atoms and show that the two atoms repel each other for $x < x_0$ and attract each other for $x > x_0$. Determine x_0.

[Ans. $\left(\dfrac{a}{b}\right)^{1/6}, \left(\dfrac{2a}{b}\right)^{1/6}, \left(\dfrac{2a}{b}\right)^{1/6}$]

CHAPTER 2	QUANTUM MECHANICS

Q.1. Electrons are accelerated to 344 Volts and are reflected from a crystal. The 1st reflection maximum occurs when glancing angle is 60°. Determine the interatomic spacing of the crystal.

[Ans. 0.38 Å]

Hints : $\lambda = \dfrac{h}{mv} = \dfrac{h}{\sqrt{2m\,eV}}$ and $d = \dfrac{n\lambda}{2\sin\theta}$

Q.2. Find the energy of the neutron in units of eV whose de-Broglie wavelength is 1 Å. Mass of neutron is 1.674×10^{-27} Kg.

[Ans. 0.0813 eV]

Q.3. A spectral line has $\lambda = 4000$ Å. Calculate frequency and the energy in eV of the photon associated with it.

[Ans. 3.1 eV]

Hints : $E = h\nu = \dfrac{hc}{\lambda}$

Q.4. In an experiment tungsten cathode which has a threshold wavelength 2300 Å is irradiated by UV light of $\lambda = 1800$ Å. Calculate (i) maximum energy of emitted photoelectrons, and (ii) work function of tungsten.

[Ans. 1.485 eV, 5.38 eV]

Hints : $T_{max} = \dfrac{1}{2}mv^2 = h(\nu - \nu_0) = h\left(\dfrac{c}{\lambda} - \dfrac{c}{\lambda_0}\right)$; $\varphi = h\nu_0$.

Q.5. Calculate the stopping potential for the photoelectrons emitted by a gold cathode if the wavelength of incident radiation is 2×10^{-7} mt ($\varphi = 4.8$ eV).

[Ans. 1.41 eV]

Hints : $T = \dfrac{hc}{\lambda} - \varphi = eV_s$

Q.6. The energy required to remove an electron for sodium is 2.3 eV. Does sodium show a photoelectric effect for orange light with $\lambda = 6800$ Å?

[Ans. 5380 Å]

Q.7. A photon of $\lambda = 3310$ Å falls on photocathode and eject an electron of energy 3×10^{-19} J. If λ is changed to 5000 Å, the energy of the ejected electron is 0.972×10^{-19} J. Calculate h, threshold frequency and φ.

[Ans. 6.62×10^{-34} J-s, 4.53×10^{14} per sec, 1.875 eV]

Hints : $T_1 = h\nu - h\nu_0$, $T_2 = h\nu_1 - h\nu_0 \Rightarrow h = \dfrac{\lambda\lambda_1 (T_1 - T_2)}{c (\lambda_1 - \lambda)}$

$\nu_0 = \nu - \dfrac{T}{h}$, $\varphi = h\nu_0$

Q.8. One millwatt of light of $\lambda = 4560$ Å is incident on Cesium surface. Calculate the photoelectric current liberated assuming a quantum efficiency of 0.5% ($\varphi = 1.93$ eV).

[Ans. 1.856 μA]

Hints : $E = \dfrac{hc}{\lambda}$; 1 mW = 10^{-3} J/s

No. of photons in 1 mW of light

$$= \frac{10^{-3}}{4.3 \times 10^{-19}} = 2.32 \times 10^{15} \text{ photons per second}$$

$Q_{eff} = 0.5\%$ No. of photoelectrons emitted from Cs surface,

$$n = \frac{0.5 \times 2.32 \times 10^{15}}{100} / \sec$$

Photoelectric current $= I = ne$

Q.9. Calculate the wavelength of thermal neutrons at 27°C assuming energy of a particle at absolute temperature T is of the order of kT (k is Boltzmann's constant and $k = 1.38 \times 10^{-23}$ J/K).

[Ans. 1.77 Å]

Hints : $\lambda = \dfrac{h}{\sqrt{mkT}}$

Q.10. Calculate the de-Broglie wavelength of an alpha particle accelerated through a potential difference of 2 KV.

[Ans. 2.3×10^{-3} Å]

Hints : $E = 2$ eV; $\lambda = \dfrac{h}{\sqrt{2m \text{ eV}}}$

Q.11. Calculate the de-Broglie wavelength of an electron in Å of energy 'V' eV.

[**Ans.** $\dfrac{12.28}{\sqrt{V}}$ Å]

Q.12. Calculate the wavelength associated with an electron subjected to a potential difference of 1.26 KV.

[**Ans.** 0.4 Å]

Q.13. Calculate the de-Broglie wavelength of a proton moving with velocity which equals to 1/20 of the velocity of light.

[**Ans.** 2.634×10^{-14} mt]

Hints : $v = c/20,\ \lambda = \dfrac{h}{mv}$

Q.14. Calculate the de-Broglie wavelength of neutron of energy 28.8 eV. Given, $m_n = 1.674 \times 10^{-27}$ kg.

[**Ans.** 0.5 Å]

Q.15. What is the energy of γ-ray photon with wavelength is 1 Å.

[**Ans.** 19.86×10^{-18} J]

Hints : $E = \dfrac{hc}{\lambda}$

Q.16. Show that the de-Broglie wavelength for a material particle of rest mass m_0 and charge q, accelerated from rest through a potential difference of V volts relativistically is given by

$$\lambda = \frac{h}{\sqrt{2m_0qV\left\{1 + \dfrac{qV}{2m_0c^2}\right\}}}$$

Hints : $T = eV,\ E^2 = p^2c^2 + m_0^2c^4,\ E = T + m_0c^2$.

Q.17. A beam of monoenergetic neutrons corresponding to 27°C is allowed to fall on a crystal. A first order reflection is observed at a glancing angle of 30°. Calculate the Interplaner spacing of the crystal.

[**Ans.** 1.78 Å]

Hints : $p = \sqrt{2mkT},\ \lambda = \dfrac{h}{p}$

CHAPTER 3 & 4 | **FREE ELECTRON THEORY AND BAND THEORY**

Q.1. An electron is confined to move between two rigid walls separated by 10^{-9} m. Find the de-Broglie wavelengths representing the first three allowed energy states of the electron and the corresponding energies?

[**Ans.** 20 Å, 10 Å, 6.7 Å, 0.37 eV, 1.48 eV, 3.33 eV]

Hints : $L = n\dfrac{\lambda}{2}$, $E_n = \dfrac{h^2 n^2}{8mL^2}$

Q.2. An electron is confined to a one dimensional potential box of side 1 Å. Obtain first two eigen values of the electron in eV?

[**Ans.** 36 eV, 144 eV]

Q.3. Determine the temperature at which there is one percent probability that a state with an energy 0.25 eV above the Fermi energy will be occupied by an electron?

Hints : $f(E) = \dfrac{1}{e^{[E - E_F]/kT} + 1}$

Q.4. Calculate the thermionic emission of a tungsten filament of length 0.05 m and area of cross section $5\pi \times 10^{-6}\,\text{m}^2$ at a temperature 2400 K? (Φ = 4.5 eV).

[**Ans.** 0.324 Å]

Hints : $I = JA = \dfrac{em}{2\pi^2 \hbar^3}(kT)^2 A$

Q.5. Copper has density and electric conductivity at 300 K as $8.96 \times 10^3\,\text{Kg/m}^3$ and $6.4 \times 10^7\,\Omega^{-1}\text{m}^{-1}$ respectively. Determine the relaxation time?

[**Ans.** 2.67×10^{-14} s]

Hints : $\sigma = ne^2\tau/m$

Q.6. Hall voltage of 1 mV is found to be developed when a sample carrying a current of 10 mA is placed in a transverse magnetic field of 3 KG. Calculate the charge carriers concentration of the sample. Given thickness of the sample along the direction of magnetic field is 0.3 mm?

[**Ans.** $6.25 \times 10^{22}\,\text{mt}^{-3}$]

Hints : $R_H = \dfrac{V_H d}{IB} = \dfrac{1}{ne}$

Q.7. An *n*-type Ge sample has a donar density of $10^{21}/\text{m}^3$. It is arranged in a Hall experiment having magnetic field of 0.5 T and current density is 500 A/m². Find the Hall voltage of the sample 3 mm wide?

[**Ans.** 4.7 mV]

CHAPTER 6	MAGNETIC MATERIALS

Q.1. The horizontal component of flux density of the earth's magnetic field is 1.7×10^3 W/m^2. What is the horizontal component of the magnetic intensity?

[**Ans.** 13.5 A/m]

Hints : $H = \dfrac{B_0}{\mu_0}$

Q.2. A bar magnet has a coercivity of 5×10^3 A/m. It is desired to demagnetize it by inserting it inside a solenoid 10 cm long and having 50 turns. What current should be sent through the solenoid?

[**Ans.** 10 A]

Hints : $H = nI$

Q.3. An iron rod of volume 10^{-4} m^3 and relative permeability 1000 is placed inside a long solenoid wound with 5 turns per cm. If a current of 0.5 A is passed through the solenoid find the magnetic moment of the rod?

[**Ans.** 4.99 A-mt^2]

Hints : $B = \mu_0 (H + M) = \mu_0 \mu_r H \Rightarrow M = (\mu_r - 1) H = (\mu_r - 1) nI$

Magnetic moment $= MV$

Q.4. An iron ring of mean circumferential length 30 cm, cross section 1 cm^2 is wound uniformly with 300 turns of wire. When a current of 0.032 A flows in the windings, the flux in the ring is 2×10^{-6} Wb. Find the flux density in the ring, the magnetic intensity and the permeability of iron?

[**Ans.** 6.25×10^{-4} Wb/A-mt^2, 32 A-turns/mt, 500]

Hints : $B = \dfrac{\phi}{A}$, $\mu = \dfrac{B}{H}$, $\mu_r = \dfrac{\mu}{\mu_0}$

Q.5. The magnetic susceptibility of medium is 9.48×10^{-9}. Calculate μ_r.

[**Ans.** $(1 + 9.48 \times 10^{-9})$]

Hints : $\mu_r = 1 + X$

Q.6. A material core has 10 turns/cm of wire wound uniformly upon it which carries a current of 2.0 A. The flux density in the material is 1.0 Wb/m^2. Calculate the magnetising force and magnetization of the material. What would be the relative permeability of the core? ($\mu_0 = 4\pi \times 10^{-7}$ Wb/m^2)

[**Ans.** 2000 A-turns/mt, 7.94×10^5 A-turns/mt, 397]

Hints : $M = \dfrac{B}{\mu_0} - H$

Q.7. A typical magnetic field obtainable with an electromagnet with iron core is ≈ 1 T. Compute the magnetic interaction energy $\mu_B \times B$ of an electron spin magnetic dipole moment with it at room temperature and show that at the approximate $\dfrac{kT}{\mu_B \times B} > 1$ is valid.

[**Ans.** $\dfrac{kT}{\mu_B} \gg 1$]

Hints : $\mu_B = \dfrac{eh}{4\pi m}$, $kT = 0.025$ eV at room temperature.

Q.8. Consider a He atom in its ground state (1s). The mean radius in the Langevin formula may be approximate by Bohr radius, $<r^2> = 0.53 \times 10^{-8}$ cm. Using $N = 27 \times 10^{23}/\text{cm}^3$ for atomic density of He gas and $e^2/mc^2 = 2.8 \times 10^{-13}$ cm, calculate the diamagnetic susceptibility of He-atom.

[**Ans.** -0.71×10^{-5}]

Hints : $X = -2 \left[\dfrac{Ne^2}{6mc^2} \right] <r^2>$

Q.9. Diamagnetic Al_2O_3 is subjected to an external magnetic field of 10^5 A/m. Evaluate magnetization and magnetic flux density in Al_2O_3 ? ($\chi = 5 \times 10^{-5}$)

[**Ans.** 0.126 Wb-mt^2]
Hints : $M = XH$, $B = \mu_0 H + M$

Q.10. The susceptibility of paramagnetic $FeCl_3$ is 3.7×10^{-3} at 27°C. What will be the value of its relative permeability μ_r at 200 K and 500 K?

[**Ans.** 5.55×10^{-3}, 2.22×10^{-3}]
Hints : $C = XT$

| CHAPTER 7 | SUPERCONDUCTIVITY |

Q.1. For a given superconductor the critical fields are 1.4×10^5 and 4.2×10^5 A/m for 14 K and 13 K respectively. Calculate the transition temperature and critical field at 4.2 K?

[**Ans.** 14.5 K, 18.9×10^5 A/mt]

Hints : $H_C = H_0 \left[1 - \left(\dfrac{T}{T_C} \right)^2 \right]$, $\dfrac{H_{C1}}{H_{C2}} = \dfrac{T_C^2 - T_1^2}{T_C^2 - T_2^2}$

Q.2   The T_c for Hg with isotopic mass 199.5 is 4.185 K. Calculate its T_c when its isotopic mass changes to 203.4?

[**Ans.** 4.14 K]

Hints : $T_{c1} M_1^{1/2} = T_{c2} M_2^{1/2}$

Q.3.   Calculate the critical current density for 1 mm diameter wire of lead at 4.2 K. Given T_C for lead is 7.18 K and $H_0 = 6.5 \times 10^4$ A/m.

[**Ans.** 1.712×10^8 A/mt^2].

Hints : $I_c = 2\pi r H_C$.

INDEX